Field Guide to
Hong Kong Mangroves

Nora Fung-yee TAM
Professor
Department of Biology and Chemistry
City University of Hong Kong

Yuk-shan WONG
Professor (Chair) of Biological Sciences
Department of Biology and Chemistry
City University of Hong Kong

漁農自然護理署
Agriculture, Fisheries
and Conservation Department

香 港 城 市 大 學 出 版 社
City University of Hong Kong Press

©2000 by Agriculture, Fisheries and Conservation Department
©2000 by City University of Hong Kong

All rights reserved. No part of this publication may be reproduced, stored in a retrieval system, or transmitted, in any form or by any means, electronic, mechanical, photocopying, recording, Internet or otherwise, without the prior written permission of the City University of Hong Kong

First published 2000
Second printing 2002
Printed in Hong Kong

ISBN 962-937-056-5

Published by
City University of Hong Kong Press
Tat Chee Avenue, Kowloon, Hong Kong

Website: http://www.cityu.edu.hk/upress
E-mail: upress@cityu.edu.hk

Contents

Foreword v
Preface vii

Introduction
Mangrove Habitat 2
Structure of Mangroves in Hong Kong 3
Function of Mangroves 11
Distribution of Mangrove Stands in Hong Kong 12

Practical Guide
How Do We Carry out Ecological Surveys in Mangroves?
    Preparation 16
    Initial Observation 18
    Where Do We Sample? 18
    What Do We Measure? 20
    Safety 21

What are the Common Ecological Techniques?
    Techniques for Abiotic Factors 23
    Quadrat Analyses 26
    Transect Analyses 31
    Soil Core Sampling 33
    Handling of Animal Specimens 33
    Data Entry and Treatment 35

How Do We Identify Mangrove Plants and Animals?
    Classification of Plants in Hong Kong Mangrove Stands 45
    Identification Key to Mangrove and Associate Plants in Hong Kong 46
    Pictures of Mangrove Plants in Hong Kong 50
    Classification of Ground-dwelling Animals
        in Hong Kong Mangrove Stands 55
    Pictures of Ground-dwelling Animals in Hong Kong 56

Appendix 1 — Distribution of 44 Mangrove Stands in Six Districts of Hong Kong 68
Appendix 2 — Possible Mangrove Stands for Students' Field Visits 76
Further Reading 83
About the Authors 85
Index 87

Foreword

Hong Kong is endowed with some wonderful natural heritage. With long, rugged and tortuous coastlines, Hong Kong has diverse and fascinating coastal areas. Mangroves are one of the most interesting habitats along the coast.

The production of the *Field Guide to Hong Kong Mangroves* will facilitate people to study these special coastal habitats. Funded by the Agriculture, Fisheries and Conservation Department, this field guide is part of a package of educational and reference materials on local mangroves. It describes the latest field techniques and systematic methods for conducting ecological surveys of mangroves in Hong Kong. It also contains useful keys and diagrams for identifying the flora and fauna found in mangroves.

I congratulate the successful production of this field guide. It will serve as a useful reference for teachers and students to study mangroves in Hong Kong. With increased knowledge and understanding on them, mangroves in Hong Kong could be conserved for the present and future generations.

Mrs. Lessie Wei, J.P.
Director of Agriculture, Fisheries and Conservation
Hong Kong Special Administrative Region

January 2000

Preface

Mangroves are diverse ecosystems that are found on the fringes of sheltered tropical and subtropical shores, regularly flooded by incoming tides but then becoming exposed when tides are gone. They serve as a link between marine and terrestrial ecosystems. This habitat has unique flora and fauna, which are in close association with its abiotic factors. Due to their distinct features, mangroves have attracted lots of curiosity and attention, and are important ecosystems for education and for scientific research. Ecological studies are now included in the curriculum of secondary schools. Many teachers bring students to mangrove habitats for ecological field work and let them appreciate the beauty of our nature. It would make their data collection and entry more useful if we provide more information on the distribution of mangrove stands in Hong Kong, the occurrence of common plants and animals, and the common ecological survey techniques.

The present book is based on the findings of a three-year (1994–1997) mangrove ecological study supported by the Agriculture, Fisheries and Conservation Department of the Hong Kong Special Administrative Region. It gives a general description of Hong Kong mangroves. The ecological techniques commonly used in field studies for analyzing mangrove plant and animal communities and their environments, the identification of plants and animals, data entry and interpretation are also clearly explained. We intend to make this book a simple and user-friendly reference for teachers, students and other interested parties who would like to carry out field studies and to appreciate the complexity and uniqueness of mangrove ecosystems. We sincerely hope that with this field guide, more ecological surveys will be conducted and long-term databases on ecological information of the local mangroves be compiled.

The authors wish to express their appreciation to all who assisted in the production of this book and in field studies, in particular, Prof. C.

Y. Lu, Xiamen University, People's Republic of China. We would also like to thank Mr. Janson K. F. Wong, Research Assistant, Dr. S. G. Cheung and his Ph.D. student, Mr. W. H. Wong at the City University of Hong Kong for their assistance in field studies and identification of animals. The authors would also like to thank Mr. Desmond K. O'Toole and Elizabeth M. O'Toole for critically reading the manuscript and providing helpful comments on the English and content of the book. Last but not least, we are grateful to the Agriculture, Fisheries and Conservation Department of the Hong Kong Special Administrative Region, especially Mr. Frank S. P. Lau, Mr. Richard P. K. Chan, Mr. Patrick C. C. Lai and Mr. T. W. Tam, for their advice, continual support and stimulating discussions during the course of the present study. Without their assistance and financial support, the book would not have been completed.

Nora Fung-yee TAM
Yuk-shan WONG
City University of Hong Kong

January 2000

Introduction

The term "mangrove" comes from a Portuguese word meaning "plants that colonize and thrive in muddy sea shores or swampy areas". Some people use the term "mangal" to describe the mangrove community, and the word "mangrove" for individual plant species in a mangrove ecosystem. In this book, the term "mangrove" refers to the complex ecosystem made up of biotic (the plant and animal communities) and abiotic factors.

Introduction

Mangrove Habitat

Mangrove is an inter-tidal wetland ecosystem fringing sheltered tropical and subtropical shores. The ecosystem is affected by tides and forms a green belt along coastlines, so it is sometimes called an "estuarine habitat". The total mangrove area in the world is around 17,000,000 ha ($\approx$180,000km^2).

Mangrove habitat receives inputs from regular tidal flushing and from freshwater streams and rivers, that influence both aquatic and terrestrial ecosystems. It is characterized by high temperature, fluctuated salinity, alternating aerobic and anaerobic conditions, periodic wet and dry soils, and unstable and shifting substratum. Mangrove plants and animals therefore face stressed and rough environments.

Mangrove communities are made up of taxonomically diverse groups of salt-tolerant plants and animals with special physiological and structural adaptations to the stressed environment. Around 60 true or exclusive mangrove plant species have been identified worldwide. Mangrove soils also harbour a large variety of micro-organisms including bacteria, algae and fungi.

Each mangrove has its own distinctive flora and fauna, and physical and chemical factors. Mangrove stands vary in significant ways geographically. Even in a small territory like Hong Kong with an area of around 1,000km^2, mangrove stands in the east differ very much from those in the west. Similarly, variations in biotic and abiotic features can also be found within one district such as Sai Kung in Hong Kong.

Mangrove stand in Lai Chi Wo ▶

Structure of Mangroves in Hong Kong

Hong Kong, located at 22°30′N and 114°10′E, represents the mangroves distributed in the South China Sea, and is part of the mangrove distribution of the Western Indo-Pacific region. Mangroves are inundated by incoming tides twice a day, with the largest tidal range at high spring tide of about 2.8m. The mean annual temperature is 23°C (ranges from 0.2°C in a cold winter to 36°C in a hot summer) and annual rainfall is around 2,214mm.

Due to the cold winter climate, mangroves in Hong Kong are relatively scattered. The number of plant species, their stature and their extents are limited compared to those in the tropical areas. Most mangrove stands have a narrow belt of mangrove plants and animals. Substrates are sandy and stony in most mangrove stands in Hong Kong. Lack of muddy substrate also limits the size and diversity of mangrove plants and animals in this territory. The extent of mangroves in Hong Kong is further restricted and disturbed by human activities, in particular, cutting, reclamation, infra-structural and urban development.

▼ Location of Hong Kong

Introduction

Plant Community

Mangrove plant communities in Hong Kong are often simple without any layering. "Vertical Zonation" (as shown by different plant species from land to sea of a mangrove stand) is often not obvious except in a relatively large stand such as at Lai Chi Wo. (see map, p. 68) No significant trend of changes in tree heights and plant species is found from land to sea. The gradients in tidal flushing and salinity do not change much vertically.

A total of 20 plant species have been identified, with 8 true or exclusive mangrove species, 5 associate or non-exclusive mangrove species, and 7 other non-mangrove species, which seem to have high conservation values.

▼ *Kandelia candel*, True Mangrove

True Mangroves (8)

1. *Kandelia candel* (L.) Druce 秋茄, 水筆仔 – Kc
2. *Aegiceras corniculatum* (Linn.) Blanco 桐花樹 – Ac
3. *Excoecaria agallocha* L. 海漆 – Ea
4. *Acanthus ilicifolius* L. 老鼠簕 – Ai
5. *Avicennia marina* (Forsk.) Vierh. 白骨壤, 海欖雌, 海茄冬 – Am
6. *Bruguiera gymnorrhiza* (L.) Poir 木欖 – Bg
7. *Lumnitzera racemosa* Willd. 欖李 – Lr
8. *Heritiera littoralis* Dryand. ex W. Ait 銀葉樹 – Hl

Associate Mangroves (5)

1. *Clerodendrum inerme* (Linn.) Gaertn. 假茉莉, 苦楮 – Ci
2. *Hibiscus tiliaceus* L. 黃槿 – Ht
3. *Cerbera manghas* L. 海芒果 – Cm
4. *Acrostichum aureum* L. 鹵蕨 – Aa
5. *Thespesia populnea* (L.) Solander ex Correa 楊葉肖槿, 繖楊, 恒春黃槿 – Tp

Other Plant Species (7)

1. *Pandanus tectorius* Sol. 露兜樹
2. *Suaeda australis* (R. Br.) Moq. 南方鹹蓬
3. *Derris trifoliata* Lour. 魚藤
4. *Limonium sinense* (Girard) Kuntze 寶血草
5. *Scaevola sericea* Vahl. 草海桐
6. *Halophila ovata* Gaudich. 圓葉喜鹽草
7. *Zostera japonica* Aschers. & Graebn. 大葉藻

The dominant species are *Kandelia candel* and *Aegiceras corniculatum*, followed by *Excoecaria agallocha*, *Avicennia marina* (a pioneer mangrove species), and *Bruguiera gymnorrhiza*. Mangrove species such as *Heritiera littoralis* and *Lumnitzera racemosa* are relatively rare and can only be found in a few mangrove stands. The average heights of the dominant trees in most mangrove stands vary from 0.7 to 7m with an average height of 1.8m. The average tree density and canopy cover are 1.4 individuals m^{-2} and 70%, respectively.

Plants have developed morphological, physiological and anatomical adaptations to cope with five major problems, comprising unstable substrata, anaerobic conditions, high salinity, establishment and desiccation. The adaptations are:

1. Viviparous reproduction and production of numerous seeds to enhance reproductive success.
2. Development of cable roots and pneumatophores (aerial roots) for gaseous exchange during periods of high tide.
3. Formation of knee joints, buttress and prop root systems for aeration at high tides and anchorage in soft mud to stabilize the plant.
4. Possession of salt glands, sclerophyllous tissues, sunken stomata, corky waterproof bark and thick waxy cuticle, hairy surface and succulent leaves to tolerate high salinity and to minimize excess water loss due to transpiration and evaporation during exposure at low tide.
5. Being an evergreen and woody plant, with high primary production, fast decomposition rate and rapid nutrient turnover.

Introduction

Distribution of Mangrove Plant Species

44 Mangrove Stands in Hong Kong	True Mangrove								Associate Mangrove				
	Kc	Ac	Ea	Ai	Am	Bg	Lr	Hl	Ci	Ht	Cm	Aa	Tp
Chek Keng 赤徑	●	●	●	●	●	●	●	●	●	●	●		
Chi Ma Wan 芝麻灣	●	●	●	●	●				●	●			
Discovery Bay 愉景灣	●	●	●	●	●	●			●				
Ha Pak Nai 下白泥	●	●		●	●				●	●			
Ho Chung 蠔涌	●	●	●	●	●	●			●	●		●	
Hoi Ha Wan 海下灣	●	●	●	●					●	●	●		
Kau Sai Chau 滘西洲	●	●	●		●	●			●	●			
Kei Ling Ha Ho 企嶺下海	●	●	●	●		●	●		●	●			
Kei Ling Ha Lo Wai 企嶺下老圍	●	●	●		●	●			●	●			●
Lai Chi Chong 荔枝莊	●	●	●	●	●	●			●	●			
Lai Chi Wo 荔枝窩	●	●	●	●	●	●			●	●	●	●	
Luk Keng 鹿頸	●	●	●	●	●	●	●		●	●		●	
Lut Chau 甩洲	●	●	●			●				●		●	
Mai Po 米埔	●	●	●		●	●			●	●			
Nai Chung 泥涌	●	●	●	●		●			●	●			
Nam Chung 南涌	●	●	●	●	●	●			●	●			●
Pak Sha Wan 白沙灣	●	●			●				●	●		●	
Pak Tam Chung 北潭涌	●	●	●	●	●	●			●	●			
Pui O Wan 貝澳灣	●	●	●	●					●				
Sai Keng 西徑	●	●	●	●	●	●	●		●				
Sai Kung Hoi 西貢海	●	●	●	●	●				●				
Sam A Tsuen/Wan 三椏村／灣	●	●	●	●	●	●			●	●			
Sam Mun Tsai 三門仔	●	●	●		●	●			●	●			
San Tau 磡頭	●	●	●	●	●	●			●	●			
Sha Tau Kok 沙頭角	●	●	●	●	●	●			●	●		●	●
Sham Chung 深涌	●	●	●		●				●	●		●	
Sham Wat 深屈	●	●	●	●	●				●	●			
Sheung Pak Nai 上白泥	●	●		●	●				●	●			
Shui Hau 水口	●	●								●			
Tai Ho Wan 大蠔灣	●	●	●	●	●	●			●	●		●	
Tai O 大澳	●	●	●	●					●				
Tai Tam 大潭	●												
Tai Tan 大灘	●	●		●	●		●		●	●			
Tai Wan 大環	●	●	●	●		●			●	●			
Tan Ka Wan 蛋家灣	●	●	●	●					●	●			
Ting Kok 汀角	●	●	●		●	●	●		●	●			
To Kwa Peng 土瓜坪	●	●	●	●	●	●			●	●	●		
Tolo Pond 吐露港	●	●	●	●	●	●			●	●			
Tsim Bei Tsui 尖鼻咀	●	●	●	●					●	●			
Tung Chung 東涌	●	●	●	●	●	●			●	●	●	●	●
Wong Chuk Wan 黃竹灣	●	●	●	●	●				●	●	●		
Wong Yi Chau 黃宜洲	●	●	●	●	●				●	●			
Yuen Long 元朗	●	●							●	●			
Yi O 二澳	●	●		●	●	●			●	●	●		
Total	44	43	40	33	29	28	14	8	40	38	20	15	9

Note: Species are arranged from dominant to rare according to their occurrence. Refer to the tables on page 4 for full name of the species.

Distribution of Non-mangrove Plant Species

44 Mangrove Stands in Hong Kong	Pt	Sa	Dt	Ls	Ss	Ho	Zj	Remarks
Chek Keng 赤徑	●							
Chi Ma Wan 芝麻灣	●							Pt: *Pandanus tectorius*
Discovery Bay 愉景灣	●	●			●			Sa: *Suaeda australis*
Ha Pak Nai 下白泥	●					●		Dt: *Derris trifoliata*
Ho Chung 蠔涌			●			●		Ls: *Limonium sinense*
Hoi Ha Wan 海下灣	●							Ss: *Scaevola sericea*
Kau Sai Chau 滘西洲			●					Ho: *Halophila ovata*
Kei Ling Ha Ho 企嶺下海	●	●		●	●			Zj: *Zostera japonica*
Kei Ling Ha Lo Wai 企嶺下老圍	●	●			●			
Lai Chi Chong 荔枝莊		●						
Lai Chi Wo 荔枝窩	●			●			●	
Luk Keng 鹿頸	●							
Lut Chau 甩洲			●					
Mai Po 米埔	●	●					●	
Nai Chung 泥涌	●			●	●			
Nam Chung 南涌	●	●	●					
Pak Sha Wan 白沙灣	●							
Pak Tam Chung 北潭涌	●							
Pui O Wan 貝澳灣	●							
Sai Keng 西徑	●							
Sai Kung Hoi 西貢海	●		●					
Sam A Tsuen/Wan 三椏村／灣	●		●					
Sam Mun Tsai 三門仔	●	●	●					
San Tau 礒頭	●	●	●	●	●	●		
Sha Tau Kok 沙頭角	●	●						
Sham Chung 深涌	●							
Sham Wat 深屈	●							
Sheung Pak Nai 上白泥				●		●		
Shui Hau 水口	●	●						
Tai Ho Wan 大蠔灣	●							
Tai O 大澳								
Tai Tam 大潭				●				
Tai Tan 大灘	●				●			
Tai Wan 大環	●	●						
Tan Ka Wan 蛋家灣	●							
Ting Kok 汀角	●	●	●					
To Kwa Peng 土瓜坪	●							
Tolo Pond 吐露港	●							
Tsim Bei Tsui 尖鼻咀	●			●				
Tung Chung 東涌	●	●						
Wong Chuk Wan 黃竹灣	●			●				
Wong Yi Chau 黃宜洲	●			●				
Yuen Long 元朗								
Yi O 二澳	●	●	●			●		
Total	35	16	14	10	7	4	3	

Note: Species are arranged from dominant to rare according to their occurrence.

Introduction

Animal Community

Mangrove ecosystems support a large diversity of animals, in particular, the ground-dwelling animals, those that feed on the forest floor. They are best seen at low tides and they feed on detritus (the dead organic matter such as fallen and decaying leaves and branches). A total of 100 different ground-dwelling animal species have been recorded, belonging to 5 Phyla and 18 Orders.

Most species belong to Phylum Mollusca, Classes Gastropoda and Bivalvia (70% of total animal species) and Phylum Arthropoda, Class Crustacea (26%). Crustacean species such as *Alpheus brevicristatus* (snapping shrimp) and *Ligia exotica* (sea slater); gastropods such as *Cerithidea djadjarinesis* and *Terebralia sulcata*, are the dominant species in Hong Kong mangrove stands.

▲ Mudskipper

Crabs, the burrowing animals, are numerous in individuals and species in Hong Kong mangroves. The fiddler crabs, *Uca* spp., and different kinds of hermit crabs are most common in mangrove ecosystems. Other than crabs, *Periophthalmus cantonensis* (mudskipper) is also dominant. The average density is 116 animals per m^2 of which 66 are crustacean and 50 are mollusks. The diversity of animal species varies between stands, ranging from 5 to 44 species per stand. On average, a mangrove stand in Hong Kong has around 24 different species of ground-dwelling animals. A table of classification is printed on p. 55. The locations of mangrove stands can be found in the Appendix.

In addition to ground-dwelling animals and burrowing animals, some invertebrates such as the herbivorous snails, *Littoraria melanostoma*, are common on the leaves, branches, trunks and prop roots of mangrove trees. Encrusting fauna dominated by bivalves, with occasional heavy infestations of barnacles can be found on the lower parts of the tree trunks.

Mangroves also provide an important habitat for a wide variety of wildlife, both aquatic and terrestrial, in particular, water birds like herons, egrets and cormorants. Mangroves are always good sites for bird watching.

Introduction

For instance, at least 350 species of birds (both resident and migrant / visiting) have been recorded at Mai Po. More than 50,000 water birds including endangered species such as Saunders' Gull and Black-faced Spoonbills visit Deep Bay every winter.

The ground-dwelling animals in Hong Kong mangrove stands are separated into three groups according to order of occurrence. These are (1) common species (can be found in most of the mangrove stands), (2) frequent species (can be found in some mangrove stands), and (3) uncommon (can only be found in one mangrove stand in Hong Kong). Their distribution is shown in the following tables.

Distribution of Common Mangrove Animal Species

Species	Chek Keng 赤徑	Ho Chung 蠔涌	Hoi Ha Wan 海下灣	Kei Ling Ha Ho 企嶺下河	Kei Ling Ha Lo Wai 企嶺下老圍	Lai Chi Wo 荔枝窩	Lut Chau 甪洲	Nam Chung 南涌	Pui O Wan 貝澳灣	Sai Keng 西貢	Sam Mun Tsai 三門仔	San Tau 新頭	Sha Tau Kok 沙頭角	Sheung Pak Nai 上白泥	Tai Ho Wan 大蠔灣	Tai Tan 大灘	Tai Wan 大環	Ting Kok 汀角	Tolo Pond 吐露池	To Kwa Peng 土瓜坪	Tsim Bei Tsui 尖鼻咀	Wong Yi Chau 黃宜洲	Yi O 二澳
Alpheus brevicristatus 短脊鼓蝦	●	●	●	●	●	●			●		●	●	●	●	●	●	●		●		●	●	
Cerithidea djadjariensis 查加擬蟹守螺	●	●	●	●	●	●			●	●	●	●	●	●	●	●	●		●	●	●	●	
Cerithidea microptera 小翼擬蟹守螺	●		●	●	●	●			●	●	●	●	●	●	●	●	●		●	●	●	●	
Cerithidea rhizophorarum 紅樹擬蟹守螺	●		●		●	●					●	●	●	●	●	●	●		●	●	●	●	
Clithon oualaniensis 奧萊彩螺	●	●	●	●	●			●	●	●	●	●	●	●	●	●	●		●	●	●		
Geloina erosa 掉地蛤	●	●	●	●	●	●				●	●	●	●	●	●	●	●		●		●	●	
Ligia exotica 海蟑螂	●	●	●	●				●	●	●	●	●	●	●	●	●	●		●		●	●	
Onchidium verruculatum 石磺	●	●	●	●	●	●	●		●	●	●	●	●	●	●	●	●		●	●	●	●	
Pagurus sp. 寄居蟹	●	●	●	●				●	●	●	●	●	●	●	●	●	●		●		●	●	
Periophthalmus cantonensis 彈塗魚	●	●	●	●	●	●			●	●	●	●	●	●	●	●	●	●	●	●	●	●	
Saccostrea cucullata 僧帽牡蠣	●	●	●	●	●			●	●	●	●	●	●	●	●	●	●		●	●	●	●	
Terebralia sulcata 溝紋筍光螺	●		●	●	●	●						●	●	●	●	●	●		●		●	●	
Uca chlorophthalmus crassipes 粗腿綠眼招潮蟹	●		●	●	●	●			●	●	●	●	●	●	●	●	●		●		●	●	
Uca vocans vocans 凹指招潮蟹	●	●		●	●	●					●	●		●			●		●		●		
Barnacle species 藤壺	●	●	●	●	●				●	●	●	●	●	●	●	●	●		●		●	●	
Oyster species 蠔、牡蠣	●	●	●	●	●				●	●	●	●	●	●	●	●	●		●		●	●	

Note: Species are arranged in alphabetical order. Locations are shown in map on pp. 68–69.

Introduction

Distribution of Frequent Mangrove Animal Species

Species	Chek Keng 赤徑	Ho Chung 蠔涌	Hoi Ha Wan 海下灣	Kei Ling Ha Ho 企嶺下海	Lai Chi Wo 荔枝窩	Lut Chau 甩洲	Nam Chung 南涌	Pui O Wan 貝澳灣	Sai Keng 西徑	Sam Mun Tsai 三門仔	San Tau 龍頭	Sha Tau Kok 沙頭角	Sheung Pak Nai 上白泥	Tai Ho Wan 大蠔灣	Tai Tan 大灘	Ting Kok 汀角	Tolo Pond 吐露湖	To Kwa Peng 土瓜坪	Tsim Bei Tsui 尖鼻咀	Wong Yi Chau 黃宜洲	Yi O 二澳
Assiminea brevicula 短擬沼螺				●								●	●								●
Balanus reticulatus 紋藤壺	●		●					●				●		●	●						
Batillaria multiformis 多形灘棲螺	●		●					●				●	●	●	●						●
Batillaria zonalis 縱帶灘棲螺																					
Brachidontes variabilis 變化短齒蛤			●				●			●			●								
Cassidula plectorematoides 絞孔胃螺	●			●			●														
Cellana testudinaria 龜嫁戚							●														
Cerithidea cingulata 珠帶擬蟹守螺	●									●		●	●								●
Cerithidea ornata 彩擬蟹守螺					●		●					●									
Clithon faba 豆彩螺	●		●	●			●							●							
Clithon retropictus 轉色彩螺																	●				
Clithon sowerbianus 多色彩螺		●												●	●						
Clypeomorus coralia ■桑椹螺	●						●							●	●						
Clypeomorus humilis 黑榴盾桑椹螺	●													●							
Diogenes edwardsii 艾氏活額寄居蟹			●	●																	●
Ellobium chinensis 中國耳螺							●										●				
Fulvia sp. 薄殼鳥蛤	●													●	●						
Gafrarium tumidum 凸加夫蛤														●	●					●	
Littoraria ardouiniana 斑肋濱螺	●																			●	
Littoraria articulata 粗糙濱螺																					
Littoraria melanostoma 黑口濱螺	●	●					●			●			●						●		●
Littoraria pallescens 淺黃濱螺		●								●				●							●
Lunella coronata granulata 粒花冠小月螺	●		●					●													
Metopograpsus latifrons 大額蟹				●								●									
Mitra sp. 筆螺			●																		
Monodonta labio 單齒螺	●		●																		●
Nassarius dealbatus 秀麗織紋螺	●		●																		
Nerita albicilla 漁舟蜒螺			●																		
Nerita chamaeleon 矮蜒螺	●													●							
Nerita lineata 黑線蜒螺																					
Nerita polita 錦蜒螺	●																				
Nerita striata 條蜒螺	●		●					●		●				●					●	●	
Nerita yoldii 齒紋蜒螺	●																				●
Neritina (Dostia) violacea 紫蜒螺				●	●					●			●	●							
Pyramidella sp. 小塔螺							●							●							
Scylla serrata 鋸緣青蟹			●									●									
Sesarma (Chiromantes) bidens 雙齒相手蟹			●															●			●
Stenothyra sp. 狹口螺																	●				
Uca(Deltuca) arcuata 弧邊招潮蟹	●			●									●						●		
Uca lactea annulipes 環紋清白招潮蟹				●															●		
Clithon species 彩螺	●			●										●						●	

Note: Species are arranged in alphabetical order.

Distribution of Uncommon Mangrove Animal Species

Species	Chek Keng 赤徑	Ho Chung 蠔涌	Hoi Ha Wan 海下灣	Kei Ling Ha Ho 企嶺下河	Kei Ling Ha Lo Wai 企嶺下老圍	Lai Chi Wo 荔枝窩	Lut Chau 甩洲	Nam Chung 南涌	Pui O Wan 貝澳灣	Sai Keng 西徑	Sam Mun Tsai 三門仔	San Tau 鱟頭	Sha Tau Kok 沙頭角	Sheung Pak Nai 上白泥	Tai Ho Wan 大蠔灣	Tai Tan 大灘	Tai Wan 大灣	Ting Kok 汀角	Tolo Pond 吐露港	To Kwa Peng 土瓜坪	Tsim Bei Tsui 尖鼻咀	Wong Yi Chau 黃宜洲	Yi O 二澳
Acmaea sp. 笠貝科														●									
Assiminea lutea japonica 琵琶擬沼螺						●																	
Cellana toreuma 嫁戚						●																	
Clibanarius infraspinatus 下齒細螯寄居蟹				●																			
Clibanarius longitarsus 長指細螯寄居蟹																						●	
Clypeomorus moniliferum ■ 桑椹螺	●																						
Glauconome chinensis 中國綠螂														●									
Helice tientsinensis 天津厚蟹				●																			
Helice wuana 伍氏厚蟹												●											
Iravadia quadrasi 方形埃列螺				●																			
Mictyris longicarpus 長腕和尚蟹				●																			
Nanosesarma (Beanium) batavicum 印尼小相手蟹											●												
Pagurus trigonocheirus 三角掌寄居蟹																							
Planaxis sulcatus 平軸螺																				●			
Pythia cecillei 賽氏女教士螺											●												
Retusa boenensis 婆羅囊螺													●										
Sesarma plicata 褶痕相手蟹						●																	
Thais luteostoma 黃口荔枝螺																	●						
Trapezium (Neotrapezium) liratum 紋斑棱蛤											●												
Turbo articulatus 節蠑螺																					●		

Note: Species are arranged in alphabetical order.

Function of Mangroves

Mangroves are vital for healthy coastal ecosystems because they are buffers between land and sea, and serve as a link between marine and terrestrial ecosystems. They represent rich and diverse living resources with high productivity.

The mangrove stand is one of the most productive ecosystems. Litter production from mangrove plants is large and turnover of litter is also rapid. Detritus from mangrove plants, mainly fallen leaves and branches, provides food and feed for an immense variety of aquatic animals, and supports complicated inshore detritus food webs.

Introduction

Mangroves also offer diverse habitats, breeding sites and feeding grounds for a large variety of coastal and marine species such as juvenile fish, crabs, shrimps, and mollusks. They are important in enhancing production from fisheries and aquaculture. They provide prime nesting and migratory sites for hundreds of bird species and wildlife, so they are important in maintaining biodiversity.

Apart from the contribution to plants and animals, mangroves protect our shorelines from erosion due to currents, waves, wind and storm. They also maintain shore stability, trap sediments to expand shore area, and sustain a natural ecological balance of aquatic ecosystems. They retain pollutants such as nitrogen, phosphorus and heavy metals from wastewater, and can serve as a natural water and wastewater treatment plant.

Mangroves are beautiful and scenic places and are good for recreation and tourism. They are important ecosystems for education and scientific research.

Distribution of Mangrove Stands in Hong Kong

There are 44 mangrove stands (see map, pp. 68–69) in the Hong Kong SAR, and they cover an area of around 270 ha in six districts:

1. Northeast New Territories (p. 70): 5 mangrove stands. Most are remote with limited access. This region is relatively less polluted or disturbed.

2. Deep Bay (p. 71): 6 mangrove stands. The largest mangrove stand is in Mai Po Nature Reserve with an area of 115 ha. Some area in this region is seriously polluted and disturbed.

3. Lantau Island (p. 72): 10 small stands. They scattered from east to west and from north to south. Some stands are facing strong pressure from infra-structural development of the Island.

4. Tai Tam (p. 73): the only mangrove stand left in Hong Kong Island. It is very disturbed.

5. Tolo Region (p. 73): 3 mangrove stands. This is the most disturbed region. Many mangroves in this region were damaged due to reclamation, construction of highways and housing estates. The existing Tolo mangrove stand will also be destroyed due to reclamation of the shore area for the development of the Hong Kong Science Park.
6. Sai Kung (pp. 74–75): 19 mangrove stands. Each stand is relatively smaller when compared to that in Deep Bay.

Mangrove stands in the Deep Bay region and some stands in Sai Kung and Lantau Island have soft, deep muddy substrates, which are sometimes almost of a fluid consistency. The soil layers are often thick, have high concentrations of organic matter and are anaerobic (black and smelly) just a few centimeters below the surface. Waters bathing these mangroves are brackish (slightly salt), ranging from saline seawater to almost fresh water. The in-flowing river waters and the continental runoff are rich in suspended matter and in nutrients; therefore, these waters are murky and fertile. The richly developed plankton adds further to the opacity of the water. The plants in these stands are often taller and develop into a thick structured community with several phyto-sociological zones.

On the other hand, there are stands with very sandy, compact, hard and shallow soil layers. The surfaces are covered with large sand particles, small stones, pebbles or even cobbles. Such substrates are less suitable for plants and only narrow belts of low growing shrubs can be found. Phyto-sociological structuring is therefore poor or lacking. The water bathing this type of mangroves is more oceanic.

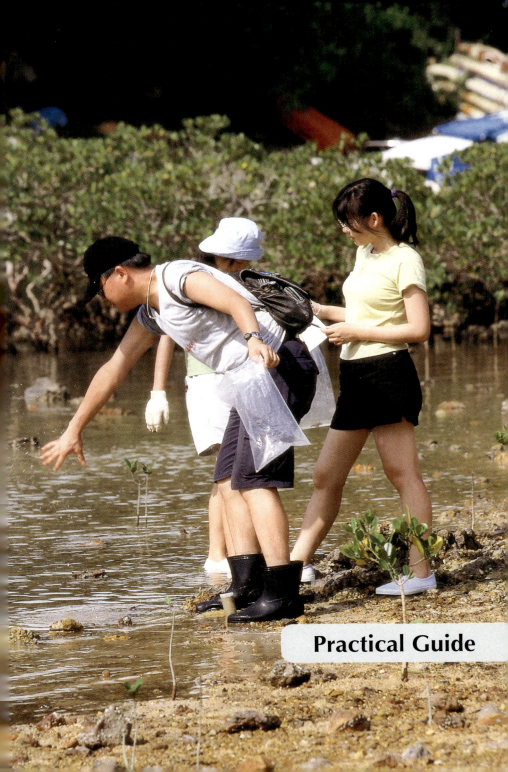

Practical Guide

How Do We Carry out Ecological Surveys in Mangroves?

Preparation

> The objectives of the field study should be clearly set and explained to students. The main objectives of any field study on mangroves can be summarized as follows:
>
> - To understand the structure, composition, and functioning of the mangrove ecosystem.
> - To appreciate the interaction and relationship between living organisms and abiotic factors, and the beauty of our nature.
> - To become familiar with common techniques for ecological studies.
> - To examine human interference and exploitation, such as pollution, influence of recreational activities, cutting of mangrove trees, collecting of mangrove animals, etc.

1. Pre-visit to the mangrove stand: teachers or school technicians should go to the selected mangrove stand 2–3 weeks before taking the students out to become familiar with the stand, and to decide the sampling position and techniques to be used. This is very important especially for stands suffering from human disturbance.
2. Collect and read information available in the literature to obtain background knowledge on mangroves, their animal and plant species.
3. Examine aerial photographs and geographical maps (1:10,000 and 1:5,000) to have a clear picture on the exact location and surroundings of the selected mangrove stand.
4. Check the weather condition from the Hong Kong Observatory, and the tidal range from the Tide Table. Inform students of the suitable time duration for doing the fieldwork, and the suitable clothing and shoes to wear. Clothes should be casual and water proof, preferably long-sleeves; shoes should be light, water proof and inexpensive. Other accessories such as cameras, insect repellant, food, water, and first aid kit should be carried.

Ecological Surveys

5. Explain to students the methodology employed, where to sample, what to measure, and what precaution to take, etc., before going out to the field. Let students become familiar with common mangrove plant and animal species. Specimens, or their photographs / pictures, from herbariums and museums would be useful tools.

6. Prepare the following items for the fieldwork:

   (a) Recording package — pencils and sign pens, checklist, collection logs, data sheets, notebook and instruction manuals. A camera will be useful.

   (b) Sampling kit
   - Knife, scissors, spades and forceps (a blunted one is very good for picking up benthos)
   - Clean collection bags and specimen bottles with labels
   - Sieves for separating animals from soil samples
   - Syringes for collecting pore water or tidal water
   - Soil cores if soil profiles need to be examined
   - Quadrats and transects
   - Other equipment for measurement of abiotic factors
   - 1:5,000 geographical map

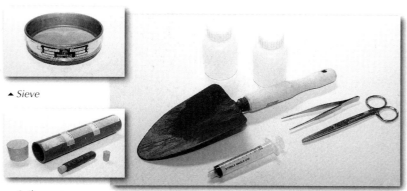

▲ Sieve

▲ Soil cores

▲ Specimen bottles, spade, forcep, scissors and syringe

Initial Observation

1. The data sheet should be ready. Date, time, and name of persons who take the readings, and location of the mangrove stand should be recorded.
2. Walk around the stand to locate the extent of the mangrove distribution and map them on a 1:5,000 geographical map.
3. Draw a rough sketch diagram of the stand, with short notes on:
   (a) General description of the mangroves — where the mangrove plants are, any woodland or wasteland or housing on the landward side, any freshwater supply (if yes, where the streams or rivers are, any variation in texture of the substrates, etc.)
   (b) Special features — any fish farms or mariculture zones nearby, any recreational activities, any illegal construction, any electrical wires, etc.
   (c) Degree of pollution or disturbance — any rubbish and what types, any wastewater discharge point, any illegal chopping of trees, any digging for clams or shells, etc.
   (d) The survey — mark clearly where the detailed survey will be carried out, and the location and orientation of transects or quadrats.
   (e) Adaptive features — observe the morphological and physiological adaptive features of plants and animals and write notes.
   (f) Photographs taken.

Where Do We Sample?

It is almost impossible to survey the whole community structure of a mangrove stand. More commonly, part of a large area is studied, subsequently generalized from the sample to the whole. Therefore, the samples must be representative of the whole. Replication of sampling units would be useful as it enhances the precision of the generalization to the whole study area.

The sample unit varies depending on what is being measured. This could be a quadrat, a belt transect, a core soil sample, a measurement of plant height, etc.

Sampling strategies — where the sample units are taken will depend on the sampling strategies which could be:

1. Random sampling: Randomness requires that each potential sample unit has an equal chance of being included, i.e., the occurrence of one sample unit will not influence the inclusion of another unit. The typical example is randomly throwing quadrats in a mangrove stand. The best way to ensure this is to give each potential sample unit a number and then choose the number using a table of random numbers.

2. Regular (Systematic) sampling: It is often convenient to take samples at regular intervals, e.g., laying quadrats along a transect. People often believe that regular sampling is advantageous because it distributes the samples all over the study area, thus ensuring that they are truly representative. It is also useful if we want to examine the changes from one side of the mangrove stand to the other end, or the vertical or horizontal distributions. However, it will produce biased results if the regularity of the sampling coincides with a natural regularity in the distribution of the organism. Also, the method is not suitable for some mangrove stands especially those that are difficult to walk through. Stratified random sampling is the solution.

3. Stratified sampling: Stratification allows separate estimates of the means and variances to be made for each stratum. Its main advantage is to allow the overall mean to be estimated with much greater precision. Stratification is also valuable if the costs of sampling are different in different parts of the study area. However, stratification may lose its advantages if the number of strata is large relative to the total number of sampling units. It is usually sufficient to use only three to six strata. This sampling strategy is less common in mangrove stands.

The exact location of sampling units and the sampling strategies will depend on the objectives of the field study, and composition and structure of the community, i.e., the heterogeneity (how varied the traits are) of the mangrove stand, as well as the degree of accuracy, precision and details desired, and time and manpower available.

What Do We Measure?

Parameters to be measured in any field study will depend on the objectives of the study. Basically two groups of parameters, abiotic and biotic factors, should be measured.

1. Abiotic factors: The distribution and abundance of plants and animals in a mangrove stand are determined to some extent by the abiotic features of the environment. They include climate, tidal range, slope of shore, nature of substrate, stone cover, soil profile (soil depth, depth of aerobic layer), and degree of anaerobiosis (redox potential, smell of H_2S, black colored zone, etc.), and pH. Measurement of these forms an integral part of most ecological studies.

2. Biotic factors: In mangrove field surveys, diversity, abundance and community structure of plants and animals (especially the ground-dwelling invertebrates) are a major focus. **Plants** are sessile, which are easy to measure. The survey can be done simply by wandering through the community and make a plant species list. The abundance of individuals within each species can be measured in terms of total counts, stem height, stem diameter, canopy size, percentage cover, number of young seedlings, etc. using quadrat or transect techniques.

   Using similar quadrat or transect techniques, the presence and absence list of **ground-dwelling animal species**, density (numbers of individuals per unit area) and biomass (wet or dry weight) are common parameters measured. Crabs, shrimps and other burrowing animals are common in mangrove stand but they are difficult to measure using quadrats. Traps and other device should be used for these fast moving and burrowing animals.

   In addition to ground-dwelling animals, **invertebrates**, due to their small size, are able to exploit very small and specific features within the environment, i.e., microhabitats. Different groups of invertebrates and at different stages of their life cycles will occupy different microhabitats, e.g., tree trunk, small branches, leaves, litter, stones, etc. This means a wide range of different microhabitats should be sampled. If time is sufficient, animals inside the mangrove soils can be observed by sampling soil cores.

Seasonal variations can be significant; therefore, it may also be necessary to sample on a number of occasions throughout the year to obtain a representative selection of species present.

Safety

Mangrove habitats are potentially hazardous and dangerous as mud may be too soft and one's feet may sink in it. Refuse including construction wastes, old furniture and electrical appliance are commonly dumped at the landward side of the mangrove stand. Wastewater from domestic, industries and agriculture may also be illegally discharged onto the mangroves. Wild dogs and dead bodies of animals may be there too. All these will cause health hazards. Therefore, safety is the most important issue to be considered before and during a mangrove field study. Anyone carrying out an ecological survey in a mangrove stand should be aware of the potential hazards and understand the safety practices.

1. No one should go and work alone in a mangrove habitat, unless there are special reasons for doing so, or work is only confined to the landward side of the stand, or someone is really familiar with the mangrove habitat.
2. Do not attempt to go to the site and work in the mangroves if the weather forecast is not good. Definitely avoid thunderstorms or typhoons.
3. Inform parents or friends about the date and time of departure, where the study site is and the expected duration of survey. Always bring a mobile phone for communication purpose.
4. Wear suitable clothing (e.g., thin long sleeve shirts are good for summer), shoes (e.g., waterproof boots with thick socks) and hat or cap. Short pants are not recommended especially if the work

needs to walk through mangrove plants. One can change clothing and shoes if necessary. Bring raincoat or waterproof clothing. Never bring umbrella. Bring sufficient water especially in hot and humid summer. If one feels sick, please inform teachers or team members. Do not over-work. Bring first aid kits.

5. Do not allow any open wound. Always cover it with waterproof plasters. Also the mangrove plants in most stands of Hong Kong are around 1–2m height with lots of dead and dried branches. These are sharp objects and anyone walking through these plants can easily get scratches. So walk carefully inside a dense patch of mangrove plants.

6. Do not carry too much. Carry only those needed for the survey work, as it is not easy to walk through a dense mangrove stand. However, do not leave anything such as transect behind.

7. As mangroves are located in inter-tidal zone, the period available for work is usually limited by the tides, knowledge of the time of the tide and the extent of tidal range is essential. Nevertheless, each day's tide may vary due to changes in weather conditions, e.g., a change to an onshore wind can bring forward the time of high tide. Therefore, allow sufficient time to do the work. Allow time to wander back from the work area to the landward side before the tidal water is too high to paddle through.

8. The mud condition of a mangrove stand varies, from stony, sandy to soft mud. When wandering around the mangroves, especially when crossing a river channel or a creek, one must be mindful of the depth and the nature of the bottom because soft and deep mud creates problems in these places. Test before each footstep. A stout pole or a stick is useful, and walking onto roots may help. Do not cling or rely too much on the branches, as some of them are dry and bristle.

When your feet sink into the mud or get stuck, do not panic and do not make violent movements, as this will make the situation worse. Gradually slip one foot out (or even leave your boot inside the mud) and rest the leg on the surface or on the roots, and gradually free the other foot will help. One can also ask your team members to pull you out but make sure their feet are on solid ground.

What are the Common Ecological Techniques?

Techniques for Abiotic Factors

1. Climatic data: These include light duration, light intensity, cloudiness, relative humidity, air temperature, wind speed and direction. All these data can be obtained from the Hong Kong Observatory.

2. Temperature of the habitats: The temperature of pore water and foreshore water, surface and at different depths of soil, below the leaf or canopy, and any other interesting microhabitats can be measured using an ordinary mercury bulb thermometer. Continuous temperature records can be made with thermographs.

3. Humidity: The relative humidity of the atmosphere and microhabitats can be determined with the dry and wet bulb psychrometer or the hydrograph.

4. Slope profile:

    (a) On any shore from Lowest Water Mark (LWM) to Highest Water Mark (HWM), there will be a major gradient of changing conditions at right angles to the water line. This varies from fully marine (below LWM) to fully terrestrial (above HWM). The LWM and HWM can be obtained from the Hong Kong Observatory as a reference.

    (b) The slope of the shore can be indicated by means of a slope profile, commonly measured by means of a device known as a *Rules and Spirit Level*. This device comprised two surveyor's poles (or 2 long sticks painted with alternating white and red color at each 10 cm intervals), a meter rule or a long stick with a spirit level lashed to it, and a tape measure (refer to the figure on p. 24). This method is direct, simple and effective over a short distance on a steep slope. However, it is less useful on gentle gradients and very inaccurate if used over long distances.

Rules and Spirit Level

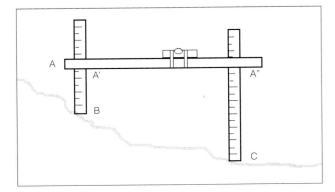

This is a simple method to measure the slope profile. The rule with a leveling device is held horizontally with one end on the known level A, then place one pole on the point B, and another pole on point C, measure A'B and A"C using a meter tape, the difference between these two values is the changes of slope level.

    (c) The *Stretch String and Rule* is an extension of the Rules and Spirit Level method in which a long cord is held straight and adjusted to a spirit level, and this cord provides the datum line. Any deviation from the datum line is measured with a meter tape. This method is very simple but requires several persons to work together. It is less convenient on steep slopes.

5. Salinity: The salinity of foreshore water or pore water can be measured by applying one drop of water sample onto the refractometer, then recording the readings from the eyepiece of the refractometer.

6. Soil data:

    (a) Soil profile — this is best studied by digging a vertical trench to get a clean and clear vertical face on one side of the trench. Record each horizon (layer) in turn down to the parent rock material. The depth of each layer, especially the difference between aerobic and anaerobic layer, should be measured.

    (b) Soil texture — this is determined according to the percentages of various particle sizes. For most biological purposes, the texture is coarsely divided into sand (2mm – 60µm), silt (60 –

2μm), and clay (< 2μm). Particle size can be measured after passing a soil sample through successive sieves of standard mesh sizes. For separating particles < 60μm, sedimentation or hydrometer methods will be employed. The former method is based on the assumption that the diameter of a particle is related to the velocity of a particle falling through a viscous medium, so soil suspension after vigorously mixing and settling will be sampled and measured. The hydrometer method measures the change in the density of soil-water suspension over time. Hand texturing is also a simple method to test the soil texture. The principle of hand texturing is that the three main size grades (sand, silt and clay) have a different feel when rubbed between the fingers. Sand feels gritty, silt feels silky and smooth, while clay is sticky and clay particles adhere to each other as well as to the skin.

(c) Stone cover percentage — the percentage of the soil surface covered by stones within a sample area (e.g., a quadrat) can be estimated visually.

(d) Soil moisture content — this is estimated by weight loss of a soil sample after drying in an oven at 105°C for 48 hours.

(e) Soil organic matter content — this is estimated by weight loss of a soil sample after ashing in a muffle furnace at >600°C for a few hours.

(f) Soil pH — the in-situ pH can be measured by inserting a pH probe directly into the soil to obtain reading. Soil pH can also be determined by making a soil-water suspension (1:10 w/v) and measuring pH with a pH meter.

(g) Soil redox potential — this is measured by inserting the probe of a redox potential meter directly into the soil and taking a reading.

(h) Amount of litter — this can be estimated by collecting all fallen and decaying leaves, branches and reproductive organs within a sampling area (e.g., a quadrat) in labelled plastic bags. The litter is then sorted in the laboratory, gently washed, dried in an oven at 70–80°C for 48 hours, then the weight is measured.

Quadrat Analyses

What is a Quadrat?

A regular shape, usually square or rectangular, of known size is called a quadrat, and is equivalent to a plot. It is the most versatile sampling method for plant and animal analyses. However, it is also the most time consuming and labour intensive technique.

What is the Size of a Quadrat?

A quadrat should be large enough to contain significant numbers of individuals but small enough that the individuals present can be separated, counted and measured without confusion.

The minimal quadrat size is determined based on the principle of the species-area curve. In this curve, the number of species is plotted against the size (or area) of the quadrat. This is based on the assumption that the number of species that can be found should increase as the size of the quadrat increases, but the number will then level off. The point at which the species number levels off is taken as the minimal area for sampling that community. The recommended quadrat size should be slightly larger than the minimal area. The procedures are summarized as follows:

1. Find the middle of a representative area to be studied.
2. Start with the smallest possible quadrat size, usually containing just one or two plant species and count the number of species present.
3. Double the size of the quadrat using the pattern shown in the figure on p. 27 and count the number of species present. Repeat the doubling and counting procedure until the number of species counted at each doubling of quadrat size levels off.
4. Stop the work when further doubling of the quadrat size gives no new species.
5. Plot the graph of species numbers against quadrat sizes. The resulting curve is the species-area curve, and the minimal area can be estimated from the curve.

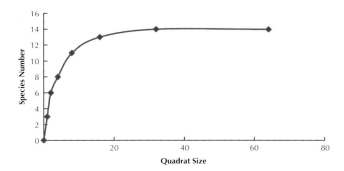

Note: Progressive doubling of quadrat size determines the minimal area of sampling or the minimal quadrat size.

Species-area Curve

Note: The Species-area Curve shows the relationship between species number and the size of quadrat (the sampling area), and determines the minimal quadrat size.

For measuring the characteristics of plants in a mangrove stand, $1m^2$ to $4m^2$ quadrats are most suitable. The quadrat sizes for ground-dwelling animals can be 0.25, 0.5 or $1m^2$.

What is the Shape of a Quadrat?

The shape of a quadrat is important in relation to the ease of laying down quadrats and to the efficiency of sampling. A quadrat can be rectangular, square or circular in shape. A long rectangular quadrat is called a belt transect.

How Many Quadrats should be Sampled?

This depends on the extent and heterogeneity of the mangrove stand, and size of the quadrat used. Whenever possible, 30 or more quadrats should be used in each mangrove stand. Any lesser number may give misleading results.

Procedures for Quadrat Analyses

After deciding the size, shape and number of quadrats to be used, quadrats should be randomly laid in the mangrove stand (random sampling) or along transect lines (regular [systematic] sampling).

For plant data, the following should be measured in each quadrat:

1. Plant species list:
   (a) Identify all the different plant species, including algae.
   (b) Plants falling on the boundary line of the quadrat should be counted only if more than one-half of the plant is within the quadrat.
2. Plant or canopy cover:
   (a) Count the number of mature individuals and young seedlings.
   (b) Estimate the coverage of plant canopy (canopy cover) or the basal area (basal cover) visually, this is called visual estimation.
   (c) Measure canopy diameter or circumference or width and length.
   (d) Measure biomass by clipping or harvesting the above ground vegetation and placing it in plastic bags. In the laboratory, sort them and determine their dry weights by drying a known

weight of sample in an oven at 70–80°C. Biomass is usually expressed in terms of grams(g) dry weight per unit area. Biomass can also be measured by a non-destructive method based on the regression equation between biomass, stem height and stem diameter.

3. Plant height.
4. Stem diameter:
   (a) Diameter of breast height (dbh) or breast height diameter — this is measured at a point 1.3m above the ground. Unless specified, dbh usually includes the thickness of the bark.
   (b) When a stem forks below breast height, or sprouts from a single base close to the ground or above it, measure each branch as a separate stem.
   (c) For small trees, stem basal diameter instead of breast height diameter can be measured. For plants with prop or buttress root systems, basal diameter should be measured just above them.

The plant data obtained from the quadrat analyses can be summarized as follows:

1. Absolute density (number of individuals of a species per m^2)
   (a) In a quadrat
   $$= \frac{\text{Number of individuals in a quadrat}}{\text{area of a quadrat}}$$
   or
   (b) In many quadrats
   $$= \frac{\text{Total number of individuals}}{(\text{total number of quadrats})(\text{area of one quadrat})}$$

2. Relative density (%)
   $$= \frac{(\text{Total number of individuals of a species})(100)}{\text{total no of individuals of all species}}$$

3. Absolute dominance (dominance per m^2)

   Dominance is expressed as basal cover, or canopy cover, biomass or weight.

(a) In a quadrat

$$= \frac{\text{Dominance of a species in a quadrat}}{\text{area of one quadrat}}$$

(b) In many quadrats

$$= \frac{\text{Total dominance of a species}}{(\text{number of quadrats})(\text{area of one quadrat})}$$

4. Relative dominance (%)

$$= \frac{(\text{Absolute dominance of a species})(100)}{\text{total absolute dominance of all species}}$$

5. Frequency (%)

$$= \frac{\text{Number of quadrats a species occurs in}}{\text{total number of quadrats}}$$

6. Relative frequency (%)

$$= \frac{(\text{Frequency of a species})(100)}{\text{total frequencies of all species}}$$

7. Relative importance (range from 0 to 300%)

$$= \text{Relative density} + \text{relative dominance} + \text{relative frequency}$$

For fauna data, the following should be measured in each quadrat:

1. Identify all animal species on the substrate.
2. Count numbers of individuals for each species and calculate density (number of individuals per m^2).
3. Collect individuals and measure their wet weight and dry weight to obtain biomass (weight per m^2).
4. Count number of crab holes to estimate numbers of crabs, as it is difficult to catch or see them in the field.
5. Draw sketch diagrams, e.g., kite diagram, based on density or biomass data to show zonation pattern if quadrats are laid along a transect.
6. Overturn stones or boulders to check for any additional animals and return stones to the original position.
7. Look for any animals on roots, stems, leaves, litters and stones, and observe their feeding behavior and movement.

Transect Analyses

What is a Transect?

A transect is a line along which all plants and / or animals are counted and measured (a line-intercept). A transect may also be a long strip of terrain in which samples of plants and / or animals are taken (a belt transect). It is a very popular approach especially for vegetation work.

Transects are usually set up deliberately across areas where there are distinct changes in vegetation and marked environmental gradients (changes of environmental factors such as temperature, moisture, light intensity through the space). It is useful to assess the distribution of plants from landward to seaward side.

A transect can also be used to determine the distribution and zonation pattern of animals present, e.g., from the upper to lower shore.

The number of transects per mangrove stand investigated varies, depending on the heterogeneity of the stand and also the objective of the study. Usually two to five transects for a mangrove stand in Hong Kong are sufficient, except for Mai Po or Lut Chau which are very large in area (see p. 71).

What is Line-intercept Analysis?

When a transect is reduced to a line, it is called a line-intercept. It is a very common method for vegetation analyses. This approach involves recording the presence or absence of plant species continuously, systematically or at random along a line marked by a tape or string.

Procedure:

1. Identify the direction and nature of the environmental gradient.
2. Lay out a line marked by a tape or a string across the zone and preferably be perpendicular to the coastline.
3. Move along the line and identify or label the species present according to one of the following three sampling methods:
    (a) Continuous sampling — record presence or absence of all species along the whole line.

(b) Systematic sampling — record presence or absence of the species at a fixed regular interval along the line.

(c) Random sampling — record presence or absence of the species at the sampling points.

4. The percentage cover of a plant species can also be estimated by measuring the length of the plant parts of each species that touch the line, or overlie or underlie the line, i.e., the intercept length. The length of intercept segments overlying bare ground may also be measured and recorded in the same manner.

5. Data collected from line-intercept analysis can be calculated as follows:

(a) Percentage cover or absolute dominance (%)
$$= \frac{\text{(Total intercept length of a species)}(100)}{\text{total length of the line}}$$

(b) Relative dominance (%)
$$= \frac{\text{(Total intercept lengths of one species)}(100)}{\text{total of intercept lengths of all species}}$$

(c) Frequency (%)
$$= \frac{\text{(Intervals in which a species occurs)}(100)}{\text{total number of intercept intervals sampled}}$$

(d) Relative frequency (%)
$$= \frac{\text{(Frequency of a species)}(100)}{\text{total frequency of all species}}$$

(e) Relative density (%)
$$= \frac{\text{(Total individuals of a species in a line)}(100)}{\text{total individuals of all species in a line}}$$

6. Distribution of plant species and the form of the plants along the line can be drawn together with the profile diagram or with the environmental data. This is a good illustration to show the relationship between slope profile (or the environmental gradient) and the zonation of plant species.

7. Add relevant environmental or land use detail.

What is a Belt Transect?

1. Where quadrats are laid out next to each other or contiguously along a transect line, the result is a belt transect.
2. Two lines that are laid parallel to each other with some distance apart, the long strip of terrain is also called a belt transect.
3. The procedures, parameters to be measured and data analysis are the same as those for quadrat analyses.

Soil Core Sampling

Core samples are collected to examine the density and diversity of animals living inside the soil or sediment, i.e., the burrowing animals. The vertical distribution of these infaunal animals in soil profile can also be determined.

Many types of core samplers can be used. The simple one is made from cylindrical metal pipe or PVC pipe (4 to 5cm in diameter), and is operated by pushing the sampler gently into the sediment by hand. Then the core sampler is retrieved from the sediment, and the sample stored in a plastic bag.

Sieve the entire core sediment in the laboratory and record the number and species of burrowing animals. The species diversity (i.e., number of different species) and the number of individuals of each species per unit area or volume of the core sampled can be calculated.

For vertical distribution, the core sediment should be separated into different layers according to its depth, e.g., top 2cm, mid-column, and bottom 2cm; or top 2cm, 2–5cm, 5–10cm, or >10cm (depending on the depth of the column and the objective of the study). Then the sediments in each depth are sieved, and the number and species recorded. The diversity and abundance of animals in different layers of the sediment core can be compared.

Handling of Animal Specimens

Once animals have been collected and brought back to the laboratory, it is necessary to treat them immediately before they die or decay.

For living specimens, one can place them in an aquarium with aerated seawater. This provides an opportunity for observing their behavior.

Once collected from the field, specimens can be preserved for later study or for future reference. A note should be made of any color markings prior to preservation as they tend to disappear rapidly after preservation. These patterns can be very valuable and they provide useful information for identification purpose.

The following chemicals can be used for preservation:

1. Buffered formalin (5%) is commonly used for fixation and preservation for animals already dead. Specimens can also be transferred to 70–80% alcohol for long-term storage after fixation in formalin for 24 hours.
2. If animals are alive, anaesthetization is required to relax and extend the animals prior to preservation. A variety of anaesthetic chemicals are available, including:

    (a) Menthol — allow animals to expand in clean seawater and scatter menthol crystals on the water surface, this takes up to 12 hours for anaesthetization.

    (b) Magnesium sulphate — place the animals in water then gradually sprinkle crystals on to the surface of the water, and leave the animals in water for some time. Alternatively, dissolve magnesium sulphate gradually in water over a period of some hours to produce a solution containing 20–30% magnesium sulphate, then put the animals in this solution. The latter method is good for treating nudibranchs and chitons.

    (c) Magnesium chloride — dissolve $MgCl_2.6H_2O$ in fresh or distilled water to produce a nearly isotonic solution containing about 7.5% magnesium chloride, then place the animals in it. This is widely used for anaesthetizing marine animals.

    (d) MS 222-Sandoz — this is the manufacturer's code name for ethyl m-aminobenzoate. It is commonly used as an anaesthetic for cold-blooded vertebrates; 0.01–0.02% aqueous solution is also used for anaesthetizing crustaceans, especially malacostracan, of all sizes and it takes from a few seconds to 10 or 15 minutes for full effect.

(e) Propylene phenoxetol — immerse the animals in clean seawater and add propylene phenoxetol (should not exceed 1% of the volume of water), a large globule of viscous compound will form at the bottom. This is commonly used as an anaesthetic for various invertebrate animals, such as mollusks and crustaceans (especially malacostracan).

Small mollusks can be air-dried, but for the larger specimens, the soft parts must be removed. This can be achieved by immersing the whole animal in boiling water and then using a bent pin or needle to extract the contents of the shell. If small parts of the animal remain in the upper whorls, these may be removed by spraying a fine jet of water into the shell aperture.

Animals such as brittle stars and feather stars, which tend to fragment rather easily, can be immersed in fresh water or refrigerated for a few hours prior to fixation.

To prepare a clean echinoid (e.g., sea urchin), the animal should be partially immersed (about two-thirds) in a mild bleach or sodium hypochorite solution. The use of bleach does not remove any natural color pattern of the animals. The spines can then be removed by scraping, scrubbing or pulling after an initial soaking in the bleach solution. The jaw apparatus and internal organs can also be removed by cutting round the soft peristome and scooping them out.

Preserved specimens should be placed in air-tight glass bottles or containers and clearly labelled, including the name of the animals, place of collection, date of collection, name of collector and name of identifier.

Data Entry and Treatment

For a general description of the mangrove stand, use geographical maps, sketch diagrams, photographs and short notes to describe the following data:

- Objectives of the study
- Geographical location of the mangrove stand
- Size (area) and extent (distribution) of mangrove plants in this stand

- Area and extent of mud flat (or foreshore) area in front of mangrove plants
- General observations and interesting features of the stand
- Pollution or disturbance from human activities

Abiotic data:

1. Prepare a slope profile
2. Data on temperature, light intensity, relative humidity, soil pH, depth, stone cover, salinity, etc. should be first recorded in the form as suggested in **Data Sheet 1** (p. 38). These factors should then be described in association with the occurrence of plants and animals.

Biota data:

1. For plant data measured using many quadrats along a transect (i.e., the belt transect technique), record the number and diversity of different plant species in each quadrat, and other plant data such as height, diameter, area, coverage or biomass in a data sheet. An example is shown in **Data Sheet 2** for one quadrat (p. 39). One data sheet is needed for one quadrat, i.e., additional copies of these data sheets are needed for more than one quadrats.

   Data from each quadrat should then be summarized in another data sheet to show the variations along the belt transect. An example is shown in **Data Sheet 3** (p. 40). If data from more than one transects are collected, use additional data sheets.

2. For animal data collected from quadrats along a transect, diversity of animal species, the number, coverage or biomass of each species should be recorded in a data sheet similar to that of the plant data. An example is shown in **Data Sheet 4** (p. 41). The changes along the transect can also be shown by entering data from each quadrat in **Data Sheet 5** (p. 42).

3. Data from line-intercept analysis can be tabulated as suggested in **Data Sheet 6** for absence and presence data (p. 43), and **Data Sheet 7** for the intercept length (p. 44).

4. Based on these data sheets, richness and abundance of plants and animals in the mangrove stand, and the dominant species can be identified. Variations along transects within the stand can be shown using appropriate figures such as line graphs, histograms, kite diagrams, bar or pie charts. Explain these variations in relation to the abiotic factors and observations made during the study.
5. Elaborate on the association between plants, animals and abiotic factors, and describe how these plants and animals adapt to the environment.

After presenting and discussing the data, check whether the objectives of the study have been achieved or not.

Compare your data with those recorded in previous years to see any changes in the features of the mangrove stand over time. If data is available, compare your data with other mangrove stands.

Data Sheet 1: For Recording Data on Abiotic Factors Collected from Quadrats

Date:
Time of Visit:
Transect Number:

Observers:
Site of Visit:

Factors	Quadrat Number					Notes
	1	2	3	4	N	
Light intensity below canopy						Light intensity:
Relative humidity						
Temperature below canopy						
Soil temperature						Ambient temperature:
On the soil surface						
Inside the soil						
Soil depth						Wind Speed:
Soil pH						
Stone cover on surface						
Salinity						
Tidal water						
Pore water						
Slope profile						
Meters along transect (ground distance)						
Relative level (rise or drop in vertical position)						
True level						
Others						

Data Sheet 2: For Recording Plant Data Collected from One Quadrat along a Belt Transect or at Random

Date:
Time of Visit:
Transect Number:
Quadrat Size:

Observers:
Site of Visit:
Transect Length and Interval of Quadrats:
Quadrat Number:

Individual Number	Species:			Species:			Species:		
	Dbh	height	canopy or area	Dbh	height	canopy or area	Dbh	height	canopy or area
1									
2									
3									
4									
5									
N									
Total									

Data Sheet 3: For Summarizing Plant Data Collected from Different Quadrats along a Belt Transect

Date:
Time of Visit:
Transect Number:
Quadrat Size:

Observers:
Site of Visit:
Transect Length and Interval of Quadrats:
Total Number of Quadrats Sampled:

Plant Species(*i*)	Units*	Quadrat Numbers					N	Total Value of Each Species
		1	2	3	4	5		

* Unit of measurement could be number of individuals, area coverage, biomass, height, diameter, etc. Each measurement should be summarized in one data sheet. Data entered into this data sheet are from the previous Data Sheet 2.

Data Sheet 4: For Recording Animal Data Collected from Quadrats along a Belt Transect

Date: Observers:
Time of Visit: Site of Visit:
Transect Number: Transect Length and Interval of Quadrats:
Quadrat Size: Total Number of Quadrats Sampled:

Quadrat Number	Species:		Species:		Species:	
	number of individuals	area covered or biomass	number of individuals	area covered or biomass	number of individuals	area covered or biomass
1						
2						
3						
4						
N						
Total						

Data Sheet 5: For Summarizing Animal Data Collected from Quadrats along a Belt Transect

Date:
Time of Visit:
Transect Number:
Quadrat Size:

Observers:
Site of Visit:
Transect Length and Interval of Quadrats:
Total Number of Quadrats Sampled:

Animal Species(i)	Units*	Quadrat Numbers					N	Total value of each species
		1	2	3	4	5		

* Unit of measurement could be number of individuals, area coverage, or biomass. Each measurement should be summarized in one data sheet. Data entered into this data sheet are from the previous Data Sheet 4.

Data Sheet 6: For Recording Absence and Presence of Plants along a Line-intercept

Date:
Time of Visit:
Line-intercept Number:
Total Number of Intercepts Measured:

Observers:
Site of Visit:
Total Line Length:
Interval of Measurement:

Plant Species	Occurrence in Interval						N
	1	2	3	4	5	...	

Data Sheet 7: For Recording the Intercept Length of Each Plant Species along a Line-intercept

Date:
Time of Visit:
Line-intercept Number:
Total Number of Intercepts measured:

Observers:
Site of Visit:
Total Line Length:
Interval of measurement:

Plant Species	Intercept Length in Each Interval (I)																	N	Total or Summation of (I)
	1	2	3	4	5														

How Do We Identify Mangrove Plants and Animals?

Classification of Plants in Hong Kong Mangrove Stands

Family	Genus	Species	Structure
True Mangrove Plants (8 species)			
Acanthaceae	Acanthus	ilicifolius	Shrub
Avicenniaceae	Avicennia	marina	Tree
Combretaceae	Lumnitzera	racemosa	Shrub/tree
Euphorbiaceae	Excoecaria	agallocha	Tree
Myrsinaceae	Aegiceras	corniculatum	Shrub
Rhizphoraceae	Bruguiera	gymnorrhiza	Tree
Rhizphoraceae	Kandelia	candel	Shrub/tree
Sterculiaceae	Heritiera	littoralis	Tree
Associate Mangrove Plants (5 species)			
Apocynaceae	Cerbera	manghas	Tree
Malvaceae	Hibiscus	tiliaceus	Tree
Malvaceae	Thespesia	populnea	Tree
Pteridaceae	Acrostichum	aureum	Fern
Verbenaceae	Clerodendrum	inerme	Tree
Other Plants (7 species)			
Chenopodiaceae	Suaeda	australis	Herb
Goodeniaceae	Scaevola	sericea	Shrub
Hydrocharitaceae	Halophila	ovata	Seagrass
Leguminosae-papilionaceae	Derris	trifoliata	Climber
Pandanaceae	Pandanus	tectorius	Tree
Plumbaginaceae	Limonium	sinense	Herb
Potamogetonaceae	Zostera	japonica	Seagrass

Identification Key to Mangrove and Associate Plants in Hong Kong

(A) Key to Hong Kong Mangrove Plants and Associates
(Based more on vegetative organs and is suitable for non-reproductive period)

1a.	Woody shrubs, small trees, or climbers	2
1b.	Low-growing herbs, grasses, grass-like or fern	16
2a.	Shrubs medium to tall; very long linear leaves with spiny margins arising in dense whorls from trunks	*Pandanus tectorius*
2b.	Shrubs, trees, and climbers lacking these features	3
3a.	Compound leaves, pinnate with 3-5 leaflets, climber	*Derris trifoliata*
3b.	Simple leaves, non-climbing and ascending	4
4a.	Leaf fleshy or succulent	5
4b.	Leaf not fleshy and not succulent	6
5a.	Leaves fleshy and erected, entire, less than 10 cm, with a rounded tip and an apical notch, bright green on both surfaces; short green petiole; small white five petaloid flowers in short axillary spikes	*Lumnitzera racemosa*
5b.	Leaves succulent but not erected, mostly crowded at the end of branchlets, 8–13cm long; white conspicuous flowers in axillary position, undulate petal margin	*Scaevola sericea*
6a.	Leaves holly-like, stiff, shiny with sharp spiny margin or markedly lobed or toothed; inflorescence terminal, flowers white or blue with purple hue	*Acanthus ilicifolius*
6b.	Leaves with entire margin, no spines	7
7a.	Leaves opposite	8
7b.	Leaves alternate	11
8a.	Without aerial roots; leaves entire, elliptical in shape, apex pointed, short stalk; flowers, three together at the ends of slender stalks from the leaf-axils, white in colour	*Clerodendrum inerme*
8b.	Aggregated aerial roots or buttress roots usually present	9
9a.	Aerial roots as pointed pneumatophores arise from underground cable roots; leaves bronze green, lower surface pale or white with numerous minute hairs; margin curling; flowers small with 4 inconspicuous petals (yellowish brown); fruit globate	*Avicennia marina*
9b.	Aerial roots forming arching loops from trunks (knee joints), buttress or prop root system obvious; propagules conspicuous; hypocotyl projecting out of fruit and hanging as droppers; viviparous	10
10a.	Leaf glossy green above, pale green or reddish beneath, up to 10cm long, apex bluntly pointed; red petiole 3–5cm long; droppers with 12–14 red calyx parts; stem and flowers also red in colour	*Bruguiera gymnorrhiza*
10b.	Leaf green, up to 5cm long, apex rounded, green petiole 1–1.5mm long; droppers with about 5 green calyx parts; flowers white; stem green	*Kandelia candel*

11a.	Plant body with white milky latex or colourless sap	12
11b.	Milky latex absent	13
12a.	Leaves entire, 4–10cm long; leaf base with 2 glands; flower unisexual, with male and female on different trees (dioecious), male flower in spike and yellow while female flower raceme	*Excoecaria agallocha*
12b.	Leaves entire, 6–27 cm long; leaf base without glands; flower bisexual, perfect, greenish white; calyx 5 lobed	*Cerbera manghas*
13a.	Lower leaf surfaces without silver scales and without hairs, leaves leathery, often notched at the tip; short petioles pink-red; stem smooth and reddish; droppers curved; flowers white and umbel	*Aegiceras corniculatum*
13b.	Plant body such as twigs, lower leaf surfaces and peduncle, covered with silvery white scales or haris	14
14a.	Large flat buttress roots; leaves stiff leathery, oblong shape; smooth dark green above but silvery white beneath; flowers small and organised in loose panicles; fruit green and brown when dry, has a ridge on the outer edge resembling a chicken comb	*Heritiera littoralis*
14b.	Without buttress roots; leaves heart-shape; flowers solitary	15
15a.	Leaves green, deeply cordate; pedicels erect; lower leaf surface and peduncle with scales; flower solitary, axillary; epicalyx falls off early; calyx cup-shape with 5 minute teeth; bushy small tree	*Thespesia populnea*
15b.	Leaves green and smooth above, lower leaf surface and peduncle with star-shape hairs, leaf apex sharp; flower solitary or cyme; epicalyx 2 whorls, outer whorl falls off early; calyx of 5 sepals to 2.5 cm long; many-branched, small or medium tree	*Hibiscus tiliaceus*
16a.	Fern (big type); leaf large, simple pinnately compound leaf; without flower; spore-propagate; sporangium yellowish brown	*Acrostichum aureum*
16b.	Grasses or sedges or herbs; with flower	17
17a.	Growing in the inter-tidal zone, near the foreshore	18
17b.	Growing at the back of the shore (upper tidal zone)	19
18a.	Leaves ovate	*Halophila* spp.
18b.	Leaves strap-like	*Zostera japonica*
19a.	Low-growing herb, erect or nearly so; ovate leaves	20
19b.	Grasses or sedges; flat, tough, linear leaves	21
20a.	Basal rosette of leaves; leaves narrow obovate, round and blunted at the tip; erect panicle of many white flowers	*Limonium sinense*
20b.	Low-growing with short, fleshy leaves; sessile flowers in leaf axils	*Suaeda australis*
21a.	Tall grass; broad leaves, featherly infloresecence	*Phragmites communis*
21b.	Short grasses; tough flat linear leaves, often creeping by stolons	22
22a.	Ligules absent; sedges; stem triangular; leaves in 3 rows; sheath intact	*Fimbristylis* spp.
22b.	Ligules present; marine grasses	23
23a.	Inflorescence a small simple spike; flowers in spring	*Zoysia* spp.
23b.	Inflorescence a compressed, spike-like panicle; flowers in late summer	*Sporobolus* spp.

(B) Key to Hong Kong Mangrove Plants and Associates
(Based more on reproductive organs)

1a.	Woody shrubs, small trees, or climbers	2
1b.	Low-growing herbs, grasses, grass-like or fern	16
2a.	Shrubs medium to tall; very long linear leaves with spiny margins arising in dense whorls from trunks	*Pandanus tectorius*
2b.	Shrubs, trees, and climbers lacking these features	3
3a.	Plants with seeds germinating while on parent trees, hypocotyl hanging or developing inner pericarp (viviparous, with droppers or propagules)	4
3b.	Plants without this feature, i.e., seeds germinate after separating from parent trees (no viviparous)	7
4a.	Hypocotyl projecting out of fruit; droppers very obvious, club or cylindrical or spindle shape	5
4b.	Hypocotyl developing inner pericarp	6
5a.	Droppers with 12–14 calyx parts, often red in colour; leaves up to 10 cm long and with acute apex (pointed tip); stem reddish	*Bruguiera gymnorrhiza*
5b.	Droppers with about 5 calyx parts, green in colour; leaves up to 5 cm long and with round apex; stem green	*Kandelia candel*
6a.	Leaves opposite and bronze green, pale or white beneath; flowers small, with 4 inconspicuous petals (yellowish brown); fruit globate	*Avicennia marina*
6b.	Leaves alternate, smooth; petioles slightly red; droppers curved; flowers white and umbel	*Aegiceras corniculatum*
7a.	Plant body with white latex or colorless sap	8
7b.	Plant body without white latex or colorless sap	9
8a.	Flower unisexual, with male and female on different trees (dioecious), male flower in spike and yellow while female flower raceme; exclude a poisonous milky fluid when stems or leaves are broken; leaves entire, 4–10cm long; leaf base with 2 glands	*Excoecaria agallocha*
8b.	Flower bisexual; leaves entire, 6–27cm long; leaf base without glands	*Cerbera manghas*
9a.	Leaves fleshy or succulent	10
9b.	Leaves not fleshy	11
10a.	Leaves fleshy and erected, with both surfaces similar, less than 10 cm long, entire; petioles green; small white five petaloid flowers	*Lumnitzera racemosa*
10b.	Leaves succulent but not erected; white conspicuous flowers in axillary position, undulate petal margin	*Scaevola sericea*
11a.	Pinnate leaves (3–5 leaflets); climbers	*Derris trifoliata*
11b.	Simple leaves; non-climbing or acandent	12
12a.	Leaves with sharp spiny margin (holly-like), opposite; inflorescence terminal, flowers white or purple	*Acanthus ilicifolius*
12b.	Leaves with margin entire	13
13a.	Twigs covered with silver scurf; unisexual flower; leaves simple, alternate; fruit green, brown when dry, 8cm wide, boat-shaped, with a marked keel, undivided	*Heritiera littoralis*
13b.	Plant body without silver scurf and not glabrous	14

14a.	Leaves heart-shape; flowers solitary	15
14b.	Leaves elliptical in shape, leaf apex pointed, entire, short stalk; flowers, three together at the ends of slender stalks from the leaf-axils, white in colour; wild erect or straggling shrub	*Clerodendrum inerme*
15a.	Lower leaf surface and peduncle with scurves; epicalyx caducous; flower solitary, axillary; bushy small tree	*Thespesia populnea*
15b.	Lower leaf surface and peduncle with stellate-hairs; epicalyx 2 whorls, outer whorl caducous; flower solitary or cyme; many-branched, small or medium tree	*Hibiscus tiliaceus*
16a.	Fern (big type); without flower; spore-propagate; sporangium yellow-brown; leaf large, simple pinnately compound leaf	*Acrostichum aureum*
16b.	Grasses or sedges or herbs; with flower	17
17a.	Growing in the intertidal zone, near the foreshore	18
17b.	Growing at the back of the shore (upper tidal zone)	19
18a.	Leaves ovate	*Halophila* spp.
18b.	Leaves strap-like	*Zostera japonica*
19a.	Low-growing herb, erect or nearly so; ovate leaves	20
19b.	Grasses or sedges; flat, tough, linear leaves	21
20a.	Basal rosette of leaves plus erect panicle of many white flowers	*Limonium sinense*
20b.	Low-growing with short, fleshy leaves; sessile flowers in leaf axils	*Suaeda australis*
21a.	Tall grass; broad leaves, featherly infloresecence	*Phragmites communis*
21b.	Short grasses; tough flat linear leaves; often creeping by stolons	22
22a.	Ligules absent; sedges; stem triangular; leaves in 3 rows; sheath intact	*Fimbristylis* spp.
22b.	Ligules present; marine grasses	23
23a.	Inflorescence a small simple spike; flowers in spring	*Zoysia* spp.
23b.	Inflorescence a compressed, spike-like panicle; flowers late summer	*Sporobolus* spp.

Pictures of Mangrove Plants in Hong Kong

True Mangrove Plants 真紅樹

Acanthus ilicifolius L.
老鼠簕

Aegiceras corniculatum
 (Linn.) Blanco
桐花樹

Avicennia marina
 (Forsk.) Vierh.
白骨壤，海欖雌，海茄冬

Bruguiera gymnorrhiza
 (L.) Poir
木欖

Excoecaria agallocha L.
海漆

Heritiera littoralis
  Dryand. ex W. Ait
銀葉樹

Kandelia candel
  (L.) Druce
秋茄，水筆仔

Lumnitzera racemosa
  Willd.
欖李

Associate Mangrove Plants 半紅樹

Acrostichum aureum L.
鹵蕨

Cerbera manghas L.
海芒果

Clerodendrum inerme
(Linn.) Gaertn.
假茉莉，苦楮

Hibiscus tiliaceus L.
黃槿

Thespesia populnea (L.) Solander ex Correa
楊葉肖槿，緻楊，恆春黃槿

Other Plants 其他植物

Derris trifoliata Lour.
魚藤

Halophila ovata Gaudich.
圓葉喜鹽草

Limonium sinense (Girard) Kuntze
寶血草

Pandanus tectorius Sol.
露兜樹

Scaevola sericea Vahl.
草海桐

Suaeda australis (R. Br.) Moq.
南方鹹蓬

Zostera japonica
　　Aschers. & Graebn.
大葉藻

Classification of Ground-dwelling Animals in Hong Kong Mangrove Stands

Crustacean
Alpheus brevicristatus (snapping shrimp)
Balanus spp. (branacle)
Clibanarius spp.
Diogenes edwardsii (hermit crab)
Helice spp.
Ligia exotica (sea slater)
Metopograpsus latifrons
Mictyris longicarpus (monk crab, soldier crab)
Pagurus spp. (hermit crab)
Scylla serrata (swimming crab)
Sesarma (Chiromantes) bidens (sesarmid crab)
Uca spp. (fiddler crab)
    Uca vocans vocans, Uca arcuata, Uca chlorophthalmus crassipes

Molluscs (gastropods)
Assiminea spp.
Batillaria spp.
    Batillaria multiformis, Batillaria zonalis
Cellana spp. (limpet)
Cerithidea spp.
    Cerithidea djadjariensis, Cerithidea microptera, Cerithidea rhizophorarum, Cerithidea cingulata, Cerithidea ornata
Clithon spp.
    Clithon oualaniensis, Clithon faba
Clypeomorus spp.
Littoraria spp. (*Littorina* spp.) (periwinkle)
    Littoraria ardouinana, Littoraria articulata, Littoraria melanostoma, Littoraria pallescens
Lunella coronata
Monodonta labio
Nerita spp.
    Nerita lineata, Nerita polita, Nerita striata, Nerita yoldii
Neritina (Dostia) violacea
Onchidium verruculatum (sea slug)
Planaxis sulcatus
Terebralia sulcata

Molluscs (bivalves)
Brachidontes variabilis (mangrove mussel)
Fulvia spp.
Gafrarium tumidum (small mangrove clam)
Geloina erosa (large mangrove clam)
Saccostrea cucullata (rock oyster)

Others
Periophthalmus cantonensis (mudskipper)

Pictures of Ground-dwelling Animals in Hong Kong

Crustacean 甲殼類

Alpheus brevicristatus
   (snapping shrimp)
短脊鼓蝦

Balanus species (branacles)
藤壺

Clibanarius sp.
   (hermit crab)
細螯寄居蟹

Helice species 厚蟹
   1. ***Helice tientsinensis***
      天津厚蟹

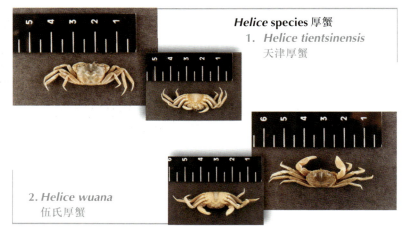

2. ***Helice wuana***
   伍氏厚蟹

Metopograpsus latifrons
大額蟹

Mictyris longicarpus
(monk crab, soldier crab)
長腕和尚蟹

Pagurus species
(hermit crab)
寄居蟹

Scylla serrata
(swimming crab)
鋸緣青蟹

Sesarma species
相手蟹

1. *Sesarma (Chiromantes bidens)* (sesarmid crab)
   雙齒相手蟹

2. Unidentified *Sesarma* Species
   相手蟹

Uca species (fiddler carbs) 招潮蟹

1. *Uca arcuata*
   弧邊招潮蟹

2. *Uca cholorophthalmus crassipes*
   粗腿綠眼招潮蟹

3. *Uca lactea annulipes*
   環紋清白招潮蟹

4. *Uca vocans vocans*
   凹指招潮蟹

Identification of Mangrove Plants and Animals

Molluscs (gastropods) 軟體動物（腹足類）

***Batillaria* species** 灘棲螺
 1. Batillaria multiformis
  多形灘棲螺

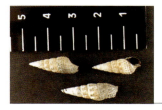

 2. Batillaria zonalis
  縱帶灘棲螺

 3. Unidentified *Batillaria* species
  灘棲螺

***Cellana* species (limpet)**
嫁䗩
 1. Cellana testudinaria
  龜嫁䗩

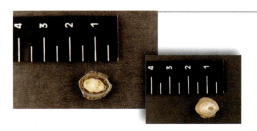

 2. Cellana toreuma
  嫁䗩

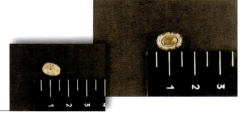

Cerithidea species
擬蟹守螺

1. *Cerithidea cingulata*
珠帶擬蟹守螺

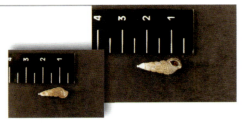

2. *Cerithidea djadjariensis*
查加擬蟹守螺

3. *Cerithidea microptera*
小翼擬蟹守螺

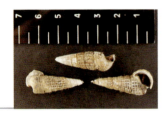

4. *Cerithidea ornata*
彩擬蟹守螺

5. *Cerithidea rhizophorarum*
紅樹擬蟹守螺

Clithon species
彩螺

1. *Clithon faba*
   豆彩螺

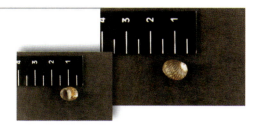

2. *Clithon oualaniensis*
   奧萊彩螺

Clypeomorus species
桑椹螺

1. *Clypeomorus humilis*
   黑瘤盾桑椹螺

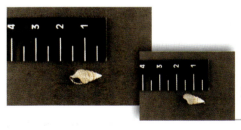

2. *Clypeomorus moniliferum*
   ■桑椹螺

***Littoraria* species (periwinkle)**
濱螺
  1. *Littoriaria ardouinana*
     斑肋濱螺

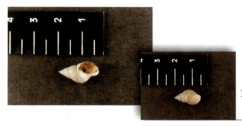

  2. *Littoriaria articulata*
     粗糙濱螺

  3. *Littoriaria melanostoma*
     黑口濱螺

Lunella coronata
粒花冠小月螺

Monodonta labio
單齒螺

Identification of Mangrove Plants and Animals

***Nerita* species**
蜑螺

1. *Nerita lineata*
   黑縫蜑螺

2. *Nerita polita*
   錦蜑螺

3. *Nerita yoldii*
   齒紋蜑螺

Neritina (Dostia) violacea
紫蜑螺

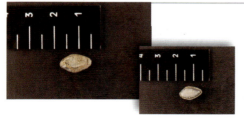

Onchidium verruculatum
  (sea slug)
石磺

Planaxis sulcatus
平軸螺

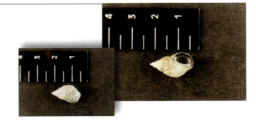

Terebralia sulcata
溝紋筍光螺

Identification of Mangrove Plants and Animals

Molluscs (bivalves) 軟體動物（雙殼類）

Brachidontes variabilis
  (mangrove mussel)
變化短齒蛤

***Fulvia* spp.**
薄殼鳥蛤

Gafrarium tumidum
  (small mangrove clam)
凸加夫蛤

Geloina erosa
  (large mangrove clam)
掉地蛤

Saccostrea cucullata
  (rock oyster)
僧帽牡蠣

Mudskipper 彈塗魚

Periophthalmus cantonensis (mudskipper)
彈塗魚

Note: In addition to crabs, shrimps, gastropods and bivalves, mudskippers (e.g., *Periophthalmus cantonensis*) are also abundant on mud-surface and mangrove branches.

Appendix

Appendix 1

Distribution of 44 Mangrove Stands in Six Districts of Hong Kong

❶ Sha Tau Kok 沙頭角

❷ Nam Chung 南涌

❸ Luk Keng 鹿頸

❹ Lai Chi Wo 荔枝窩

❺ Sam A Tsuen/Wan 三椏村／灣

❻ Mai Po 米埔

❼ Lut Chau 甩洲

❽ Yuen Long Ind. Est. 元朗工業村

❾ Tsim Bei Tsui 尖鼻咀

❿ Sheung Pak Nai 上白泥

⓫ Ha Pak Nai 下白泥

⓬ Discovery Bay 愉景灣

⓭ Tai Ho Wan 大蠔灣

⓮ Tung Chung 東涌

⓯ San Tau 䃟頭

⓰ Sham Wat 深屈

⓱ Tai O 大澳

⓲ Yi O 二澳

⓳ Shui Hau 水口

⓴ Pui O Wan 貝澳灣

㉑ Chi Ma Wan 芝麻灣

㉒ Tai Tam 大潭

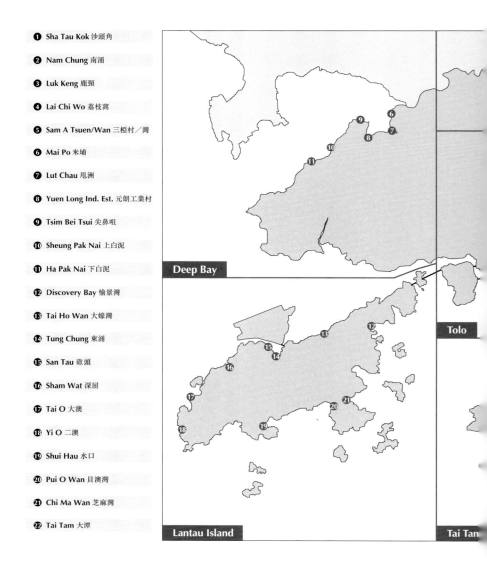

Appendix 1

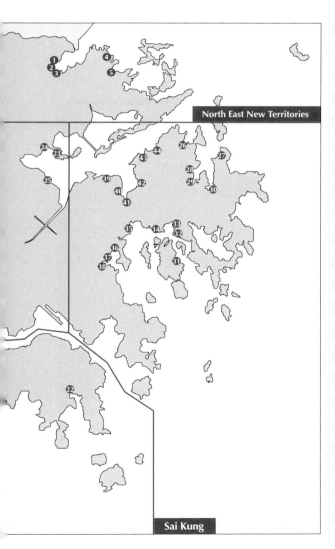

㉓ Sam Mun Tsai 三門仔
㉔ Ting Kok 汀角
㉕ Tolo Pond 吐露港
㉖ Hoi Ha Wan 海下灣
㉗ Tan Ka Wan 蛋家灣
㉘ Tai Tan 大灘
㉙ To Kwa Peng 土瓜坪
㉚ Chek Keng 赤徑
㉛ Kau Sai Chau 滘西洲
㉜ Wong Yi Chau 黃宜州
㉝ Pak Tam Chung 北潭涌
㉞ Wong Chuk Wan 黃竹灣
㉟ Tai Wan 大環
㊱ Sai Kung Hoi 西貢海
㊲ Pak Sha Wan 白沙灣
㊳ Ho Chung 蠔涌
㊴ Nai Chung 泥涌
㊵ Sai Keng 西徑
㊶ Kei Ling Ha Lo Wai 企嶺下老圍
㊷ Kei Ling Ha Hoi 企嶺下海
㊸ Sham Chung 深涌
㊹ Lai Chi Chong 荔枝莊

Appendix 1

Appendix 1

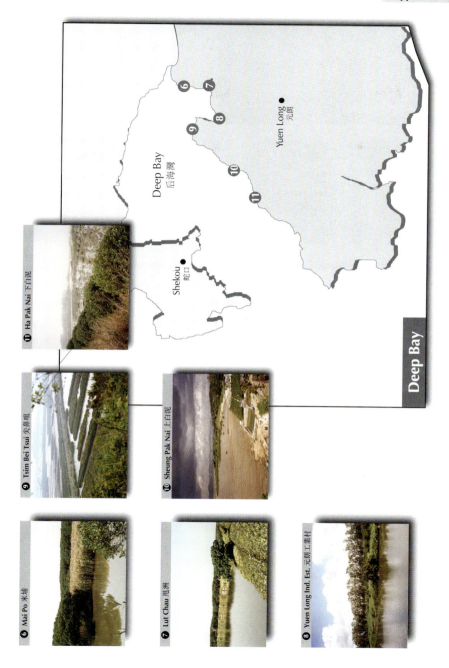

Appendix 1

Appendix 1

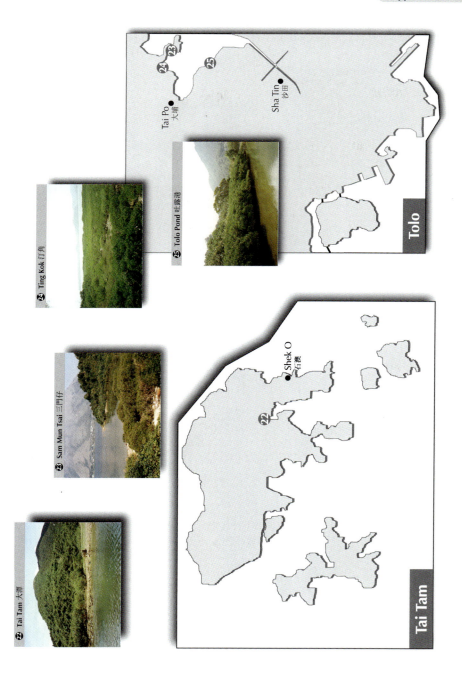

Appendix 1

Appendix 1

Appendix 2

Possible Mangrove Stands for Students' Field Visits

Each mangrove stand in Hong Kong has a different size and different ecological characteristics. Some are more complicated in terms of plant and animal community structures. Such stands require more time and effort to be spent. Some are soft mud with dense plant growth. It is difficult to walk through or to carry out any field measurement in them. Some are remote and difficult to access by public transportation although they are more natural. Therefore, the following considerations need to be taken into account when deciding which stand is suitable for students' field work:

1. Accessibility of the stand: Can the stand be accessed by public transportation? How far is it from the main road? How long does it take to travel to the stand?

2. Size of the stand: Is it large enough for the whole group of students to spread around? Is it too large or too spread so that it becomes difficult for teachers to manage? Is time sufficient for students to complete the work during the suitable tidal period? The size and extent of the mangrove stand chosen should be proportional to the numbers of students participating in the field study.

3. Ecological value of the stand: What is the species diversity, naturalness and representativeness of the stand? Has it been too disturbed by humans? Is it seriously polluted? Are the plants healthy? Can students easily find different animals?

4. Safety: Is the mud too soft and too deep for students to step onto? Are the plants too dense and difficult to walk through? Is there any broken glass or sharp rubbish? Any other hazards associated with the stand? Safety should be the most important factor for consideration.

In consideration of the points mentioned above, the following mangrove stands appear to be suitable for ecological study by senior students in secondary schools. The polluted stands are included especially for students who would like to examine the effects of pollution on mangrove ecology.

Sai Kung District

㊵ Sai Keng

It consists of 13 plant species, dominated by *Kandelia candel* and *Aegiceras corniculatum* with a few irregularly spaced *Avicennia marina*. A total of 44 species of benthic macrofauna has been identified and they are dominated by crabs and gastropods, in particular, *Cerithidea* spp. Avifauna at the stand is numerous with up to 120 egrets (*Egretta garzetta*) being counted feeding and roosting here. The stand

▲ *Sai Keng*
西徑

is part of the Kei Ling Ha Mangal SSSI (Site of Special Scientific Interests). However, dumping of building and construction refuse, and encroachments on the mangroves from extensive construction work in the village of Sai Keng have started to damage the mangroves close to the village. Pollution at the stand is minimal and restricted mainly to litter. During low tides, local villagers come and collect seafood such as clams from the foreshore mudflat. Before the building of houses in this area and the construction work, the stand was popular with families at weekends for barbecues and rowing. Some schools and colleges also use this mangrove stand for ecology fieldwork, indicating that Sai Keng has high educational and scientific value.

㊶ Kei Ling Ha Lo Wai

It consists of 11 plant species, including the locally uncommon non-mangrove plant species *Thespesia populnea*. The stand is co-dominated by a mixture of *Aegiceras corniculatum* and *Kandelia candel*. The trees are very uniform in both size and shape with an average height of 1.98m. Diversity of benthic macrofauna is low with only 23 species being recorded. The stand is dominated by crabs, in

▲ *Kei Ling Ha Lo Wai*
企嶺下老圍

particular, *Uca* spp. on the foreshore and *Sesarma* (*Chiromantes*) spp. inside the mangroves. The stand is protected by its designation as an SSSI although encroachment on the mangroves has occurred by dumping of building and construction wastes. Big stones and rubble are found at the landward side of the stand. Local village people also try to use the landward side of the mangrove stand as extra car parking places. Little pollution is in evidence though some rubbish is being dumped at the rear of the mangrove stand. The area is used for some water sports especially recreational fishing, snorkelling and rowing.

㉞ Wong Chuk Wan

It is a small stand (0.26 ha) but it spreads over around 1.5 km along the coastline (from Sai Kung, just past Tai Wan villages, to Pak Tam Chung). The stand could be considered as a composite site, consisting of many small mangrove patches concentrated at the mouths of numerous streams but each patch with low plant diversity. The community structure of this stand as a whole is very similar to that in Tai Wan and Wong Yi Chau. It is a stand suitable for students to walk around and understand the distribution of mangroves along our coastlines.

▲ *Wong Chuk Wan*
黃竹灣

㊳ Ho Chung

This stand has the freshwater Ho Chung River running through it. A total of 11 plant species is present, including the locally uncommon sea-grass species *Halophila ovata*. The stand is completely dominated by *Kandelia candel* with only occasional presence of *Aegiceras corniculatum*. The trees on the landward side of the stand are tall (>6 m) with the height decreasing slowly towards the seaward side. Benthic macrofauna is sparse and not very diverse, with only 16 species being recorded. The stand is dominated by *Cerithidea* spp. Avifauna at the stand is good with both egrets (*Egretta garzetta*) and herons being commonly found. Kingfishers (*Alcedo atthis*) and a rail (species unknown but possibly the white breasted waterhen, *Amaurornis phoenicurus*) have also been recorded at the stand. The stand is under some pressure from cutting though only a small part of the mangroves has been damaged in this fashion. The stand does not receive any protection. Pollution enters the stand from both the Ho Chung River and from a sewage discharge pipe. A lot of rubbish has been observed at the landward side of the stand where some effluent discharge and waste dumping also occurs. Despite the pollution problem, the stand is still quite healthy.

▲ *Ho Chung*
蠔涌

Appendix 2

Tolo Region

㉔ Ting Kok

This is the fourth largest mangrove stand in Hong Kong in which there are 13 plant species. Ting Kok is dominated by *Kandelia candel* and *Aegiceras corniculatum*, with a few irregularly spaced *Avicennia marina* in the middle part of the stand. Other species such as *Suaeda australis* and *Limonium sinense* are also found in this stand. The trees in the landward side of the stand are dwarfed (average height of <1.

▲ Ting Kok
汀角

2m), possibly due to the stony substrate, and the height increases to about 2.5m at the muddy seaward side. The stand has high fauna diversity, and 39 benthic species have been identified. *Cerithidea* spp., *Batillaria* spp. and *Terebralia sulcata* are the dominant species. Most of the stand lies within Ting Kok SSSI, however, there are a number of applications for development on the landward side and adjacent areas. Very little litter was noted at the stand though runoff from agricultural holdings containing contaminants and residues from fertilisers and pesticides flow into this stand. Heavy algal blooms occur in this area during summer, possibly due to eutrophication of Tolo Harbour. The stand is heavily utilised for shellfish collection and recreational fishing. Schools and universities use the stand for both teaching and research purposes. Regular field visits to the stand are frequently arranged, indicating the educational value of Ting Kok stand.

㉓ Sam Mun Tsai

The Sam Mun Tsai stand is bordered by the sea on two sides of the mangroves, and has three extremely polluted freshwater streams running into it. Twelve plant species have been recorded, dominated by *Avicennia marina*, with patches of *Kandelia candel* and *Aegiceras corniculatum*. The stand has 31 species of benthic and arboreal species, dominated by cerithid gastropods (especially *Cerithidea*

▲ Sam Mun Tsai
三門仔

rhizophorarum) and the ubiquitous *Terebralia sulcata*. The stand has not been protected and had previously been developed into a soccer pitch. But this is no longer in use and is being slowly reclaimed by semi-mangrove vegetation. The nearby egretries were designated as SSSIs by the Hong Kong SAR Government. The stand is under pressure from litter dumping. Pollution at the stand consists mainly of tide-borne flotsam. Large areas of the stand are covered with clam and mussel shells.

Lantau Island

⑱ Yi O

It has 14 plant species, co-dominated by *Avicennia marina* and *Aegiceras corniculatum*, with patches of *Kandelia candel* on the muddy substrate. The trees are dwarfed throughout the stand, with an average height of only 0.86m. A total of 40 species of benthic macrofauna has been recorded. *Terebralia sulcata* and *Cerithidea* spp. are as common as the hermit crabs (*Pagurus* spp.). The mudskipper *Periophthalmus cantonensis* is also common on the banks of the river. Yi O is not protected by any legislation although it lies on the Lantau Trail. The area is passed through frequently by walkers, especially at weekends. Pollution is minimal but tidal litter is very much in evidence especially at the high water mark and among the mangrove plants. During one visit to this stand, a lot of oil was found to be coating the flotsam and a dark layer covered the substrate.

▲ Yi O
二澳

⑮ San Tau

This stand has 18 plant species, with *Bruguiera gymnorrhiza* occurring in substantial numbers. The foreshore has patches of two locally uncommon seagrass species, *Zostera japonica* and *Halophila ovata*. The trees at the landward side of the stand are tall (average height of 2.3m). The heights of *B. gymnorrhiza* and *Excoecaria agallocha* and other trees decrease from landward to seaward side. Many dwarfed trees are found as you move towards the sea. Benthic macrofauna consists mostly of gastropods, though large numbers of hermit crabs are also present. A total of 25 species of benthic macrofauna has been recorded and the most common one is *Cerithidea djadjariensis*. Many empty gastropod shells have been found at San Tau. This stand is opposite to the new airport at Chek Lap Kok. The reclaimed area of the new airport is now cut off from the sea and a channel has been left open in the causeway to allow tidal flushing to San Tau mangroves. The effect of this is unknown. A lot of litter in the form of plastic packaging and building debris is present on the seashore and in the mangroves. The local villagers also use the back-mangroves as a convenient dumping ground for their domestic waste despite having an incinerator nearby. The stand is not heavily used for recreation but limited commercial netting is still practised. Although the stand at San Tau lies within San Tau Beach SSSI, more protection to this unique stand is needed.

▲ San Tau
礉頭

Deep Bay

⓿ Sheung Pak Nai

This stand spreads over 4km of coastline and has an area of 6.34ha. There are many polluted freshwater streams running through the mangroves into the sea. The substrate is made up of extremely soft and deep mud (very fluid) especially near the foreshore regions. Nine plant species have been recorded, including the locally uncommon species *Halophila ovata*. The stand is dominated by *Kandelia candel*, which was

▲ *Sheung Pak Nai*
上白泥

planted about 15 years ago by four local farmers to protect their fishponds from erosion. These artificially planted trees have grown very well throughout the years (some trees reach a height of ~4m) and are extremely healthy and producing large healthy propagules (or droppers). A total of 19 species of benthic macrofauna has been recorded, dominated by cerithid gastropods, mud crabs (*Helice* spp.) and mudskippers (*Periophthalmus cantonensis*). Chinese pond herons (*Ardea baccus*), grey herons (*Ardea cinerea*), night herons (*Nycticorax nycticorax*) and egrets (*Egretta garzetta*) have been recorded. The stand's proximity to Mai Po and the Deep Bay mudflats indicates more bird species (especially waders) would be found in this stand. Sheung Pak Nai lies within Pak Nai SSSI. The area is not subject to any construction or reclamation work, although offshore in this area are the biggest oyster beds in Hong Kong. These stretch all the way to Lau Fau Shan where most of the catch is landed.

Further Reading

Books, Articles and CD-ROM

Aksornkoae, S., Maxwell G. S., Havanond S. and Panicksuko S. (1992). *Plants in Mangroves*. Bangkok: Chalongrat Co. Ltd.

Baker, J. M. and Wolff W. J. (1987). *Biological Surveys of Estuaries and Coasts*. Cambridge: Cambridge University Press.

Clough, B. F. (1993). *The Economic and Environmental Values of Mangrove Forests and their Present State of Conservation in the South-east Asia / Pacific Region*. Okinawa: International Society for Mangrove Ecosystem.

Field, C. D. (1995). *Journey Amongst Mangroves*. Okinawa: International Society for Mangrove Ecosystem.

Hodgkiss, I. J. (1986). "Aspects of Mangrove Ecology in Hong Kong," in *Memoirs of the Hong Kong Natural History Society*, Hodgkiss I. J. (ed.) Hong Kong: The Hong Kong Natural History Society, No. 7, pp. 107–116.

Hodgkiss, I. J., Thrower S. L. and Man S. H. (1981). *An Introduction to Ecology of Hong Kong*, Vols. 1 & 2. Hong Kong: Federal Publications (HK) Ltd.

Holme, N. A. and McIntyre A. D. (1984). *Methods for the Study of Marine Benthos*. Oxford: Blackwell Scientific Publications.

Kent, M. and Coker P. (1992). *Vegetation Description and Analysis: A Practical Approach*. London: Belhaven Press.

Lin, P. (1988). *Mangrove Vegetation*. Beijing: China Ocean Press.

Lincoln, R. J. and Sheals J. G. (1979). *Invertebrate Animals: Collection and Preservation*. Bristish Museum (Natural History). Cambridge: Cambridge University Press.

Morton, B. and Morton J. (1983). *The Seashore Ecology of Hong Kong*. Hong Kong: Hong Kong University Press.

Semins, F. E. and Salck R. D. (1982). *Fundamentals of Ecology Laboratory Manual*. Iowa: Kendall / Hunt Publishing Company.

Snedaker, S. C. and Snedaker J. G. (1984). *The Mangrove Ecosystem: Research Methods*. Paris: UNESCO.

Sutherland, W. J. (1996). *Ecological Census Techniques: A Handbook*. Cambridge: Cambridge University Press.

Tam, N. F. Y. and Wong Y. S. (1997). *Ecological Study on Mangrove Stands in Hong Kong*. Report submitted to Agriculture and Fisheries Department, Hong Kong SAR, 5 volumes.

_____ (2000). *CD-ROM on Hong Kong Mangroves*. Hong Kong: City University of Hong Kong.

_____ (2000). *Hong Kong Mangroves*. Hong Kong: City University of Hong Kong Press.

Tam, N. F. Y., Wong Y. S., Lu C. Y. and Berry R. (1997). "Mapping and Characterization of Mangrove Plant Communities in Hong Kong," *Hydrobiologia*, 352: 25–37.

Thrower, S. L. (ed.) (1975). *The Vegetation of Hong Kong, its Structure and Change.* Symposium Proceedings of Hong Kong Branch of the Royal Asiatic Society.

Tomlinson, P. B. (1986). *The Botany of Mangroves.* Cambridge: Cambridge University Press.

Williams, G. (1987). *Techniques and Fieldwork in Ecology.* London: Bell and Hyman Limited.

Web Sites

http://life.bio.sunysb.edu/marinebio/mangal.html

http://www.agri-aqua.ait.ac.th/mangroves/ecology.html

http://www.aims.gov.au/index.html

http://www.erin.gov.au:80/sea/seascapes/mangroves/mangr_index.html

http://www.fiu.edu/~cpadil01

http://www.floridaplants.com/horticulture/mangrove.htm

http://www.floridaplants.com/mangrove.htm

http://www.iucn.org/themes/ramsar/about_mangrove_project.htm

http://www.mangrove.org/

http://www.nhmi.org/mangroves/index.htm

http://www.reefrelief.org/document/mangrove.html

http://www.zbindustries.com/mangrove.htm

The Authors

Professor Nora Fung-Yee Tam, Ph.D. did her B.Sc. (Hons.) and M.Phil. degrees at the Chinese University of Hong Kong in the late 1970s. She then went to United Kingdom for further study and obtained her doctorate in environmental biology from University of York in 1982. Since then she has taught at the then Hong Kong Polytechnic University before joining the then City University of Hong Kong in 1992. Prof. Tam is currently Professor in the Department of Biology and Chemistry at CityU. She is also appointed as Honorary Professor at Zhongshan University (Guangzhou), Xiamen University (Xiamen) and Nanjing University (Nanjing), the People's Republic of China. Professor Tam has been working on projects involving mangroves since 1986 in addition to her other research interests in algal biotechnology and waste treatment. Her mangrove projects cover areas on ecological survey of mangrove ecosystems, nutrient dynamics, plant productivity, litter decomposition, plant and animal diversity, soil properties, replanting, mangroves as wetlands for wastewater treatment, and conservation of mangroves. Prof. Tam is currently the governor of the Friends of the Earth, member of the Environment Conservation Fund Main Committee and Vetting Sub-committee, Wetland Advisory Committee of Hong Kong.

Professor Yuk-Shan Wong, Ph.D. obtained his doctoral degree in biology from McGill University of Canada in 1979. He was then awarded for a Best Scholarship and Medical Council Research Fellowship to do his post-doc at the University of Toronto. After many years of teaching and research at the Chinese University of Hong Kong, the then Hong Kong Polytechnic University, and the Hong Kong University of Science and Technology, Prof. Wong is currently Vice-President and Professor (Chair) of Biological Sciences at City University of Hong Kong. Prof. Wong's current research includes biochemical study of heavy metal tolerance mechanism in plants, algal biotechnology and mangrove study. Professor Wong is now a member of the Marine and Country Park Board and the Chairman of the Wetland Advisory Committee of Hong Kong. He also holds appointments as Adjunct Professor at Zhongshan University (Guangzhou), Xiamen University (Xiamen), and Nanjing University (Nanjing), the People's Republic of China.

Index

Mangrove Plants

Acanthus ilicifolius L. 4, 6, 45, 46, 48, 50
Acrostichum aureum L. 4, 6, 45, 47, 49, 52
Aegiceras corniculatum (Linn.) Blanco 4–6, 45, 47, 48, 50, 77, 78, 79, 80
Avicennia marina (Forsk.) Vierh. 4–6, 45, 46, 48, 50, 77, 79, 80
Bruguiera gymnorrhiza (L.) Poir 4–6, 45, 46, 48, 50, 80, 81
Cerbera manghas L. 4, 6, 45, 47, 48, 52
Clerodendrum inerme (Linn.) Gaertn. 4, 6, 45, 46, 49, 52
Derris trifoliata Lour. 4, 7, 45, 46, 48, 53
Excoecaria agallocha L. 4–6, 45, 47, 48, 51, 80, 81
Halophila ovata Gaudich. 4, 7, 45, 47, 49, 53, 78, 80, 81
Heritiera littoralis Dryand. ex W. Ait 4–6, 45, 47, 48, 51
Hibiscus tiliaceus L. 4, 6, 45, 47, 49, 52
Kandelia candel (L.) Druce 4–6, 45, 46, 48, 51, 77, 78, 79, 80, 81
Limonium sinense (Girard) Kuntze 4, 7, 45, 46, 47, 48, 53, 79
Lumnitzera racemosa Willd. 4–6, 45, 48, 51
Pandanus tectorius Sol. 4, 7, 45, 46, 48, 54
Scaevola sericea Vahl. 4, 7, 45, 46, 48, 54
Suaeda australis (R. Br.) Moq. 4, 7, 45, 47, 49, 54, 79
Thespesia populnea (L.) Solander ex Correa 4, 6, 45, 47, 49, 53, 77
Zostera japonica Aschers. & Graebn. 4, 7, 45, 47, 49, 54, 80, 81

Mangrove Animals

Acmaea sp. 11
Alpheus brevicristatus 8, 9, 55, 56
Assiminea brevicula 10
Assiminea lutea japonica 11
Balanus reticulatus 10
Barnicle species (unidentified) 9
Batillaria multiformis 10, 55, 59
Batillaria zonalis 10, 55, 59
Brachidontes variabilis 10, 55, 65
Cassidula plectorematoides 10
Cellana testudinaria 10, 59
Cellana toreuma 11, 59
Cerithidea cingulata 10, 55, 60
Cerithidea djadjariensis 8, 9, 55, 60, 80
Cerithidea microptera 9, 55, 60
Cerithidea ornata 10, 55, 60
Cerithidea rhizophorarum 9, 55, 60, 79
Clibanarius infraspinatus 11
Clibanarius longitarsus 11
Clibanarius spp. 55
Clithon species (unidentified) 10
Clithon faba 10, 55, 61
Clithon oualaniensis 9, 55, 61
Clithon retropictus 10
Clithon sowerbianus 10
Clypeomorus coralia 10
Clypeomorus humilis 10, 61
Clypeomorus moniliferum 11, 61
Diogenes edwardsii 10, 55
Ellobium chinensis 10
Fulvia sp. 10, 55, 65
Gafrarium tumidum 10, 55, 65
Geloina erosa 9, 55, 65
Glauconome chinensis 11
Helice tientsinensis 11, 56
Helice wuana 11, 56
Iravadia quadrasi 11
Ligia exotica 8, 9, 55

87

Mangrove Animals (continued)

Littoraria ardouiniana   10, 55, 62
Littoraria articulata   10, 55, 62
Littoraria melanostoma   8, 10, 55, 62
Littoraria pallescens   10, 55
Lunella coronata granulata   10, 55, 62
Metopograpsus latifrons   10, 55, 57
Mictyris longicarpus   11, 55, 57
Mitra sp.   10
Monodonta labio   10, 55, 62
Nanosesarma (Beanium) batavicum 11
Nassarius dealbatus   10
Nerita albicilla   10
Nerita chamaeleon   10
Nerita lineata   55, 63
Nerita polita   10, 55, 63
Nerita striata   10, 55
Nerita yoldii   10, 55, 63
Neritina (Dostia) violacea   10, 55, 63
Onchidium verruculatum   9, 55, 63
Oyster species (unidentified) 9
Pagurus sp.   9, 55, 57
Pagurus trigonocheirus   11
Periophthalmus cantonensis   8, 9, 55, 66, 80, 81
Planaxis sulcatus   11, 55, 64
Pyramidella sp.   10
Pythia cecillei   11
Retusa boenensis   11
Saccostrea cucullata   9, 55, 65
Scylla serrata   10, 55, 57
Sesarma (Chiromantes) bidens   10, 55, 58, 77
Sesarma plicata   11
Stenothyra sp.   10
Terebralia sulcata   8, 9, 55, 64, 79, 80
Thais luteostoma   11
Trapezium (Neotrapezium) liratum   11
Turbo articulatus   11
Uca (Deltuca) arcuata   10, 55, 58
Uca chlorophthalmus crassipes   9, 55, 58
Uca lactea annulipes   10, 58
Uca vocans vocans   9, 55, 58

Mangrove Stands

Chek Keng   6, 7, 9, 10, 11, 69, 74
Chi Ma Wan   6, 7, 68, 72
Discovery Bay   6, 7, 68, 72
Ha Pak Nai   6, 7, 68, 71
Ho Chung   6, 7, 9, 10, 11, 69, 75, 78
Hoi Ha Wan   6, 7, 9, 10, 11, 69, 74
Kau Sai Chau   6, 7, 69, 74
Kei Ling Ha Hoi   6, 7, 9, 10, 11, 69, 75
Kei Ling Ha Lo Wai   6, 7, 9, 10, 11, 69, 75, 77
Lai Chi Chong   6, 7, 69, 75
Lai Chi Wo   2, 4, 6, 7, 9, 10, 11, 68, 70
Luk Keng   6, 7, 68, 70
Lut Chau   6, 7, 9, 10, 11, 31, 68, 71
Mai Po   6, 7, 9, 12, 31, 68, 71, 81
Nai Chung   6, 7, 69, 75
Nam Chung   6, 7, 9, 10, 11, 68, 70
Pak Sha Wan   6, 7, 69, 75
Pak Tam Chung   6, 7, 69, 75, 78
Pui O Wan   6, 7, 9, 10, 11, 68, 72
Sai Keng   6, 7, 9, 10, 11, 69, 75, 77
Sai Kung Hoi   6, 7, 69, 75
Sam A Tsuen/Wan   6, 7, 68, 70
Sam Mun Tsai   6, 7, 9, 10, 11, 69, 73, 79
San Tau   6, 7, 9, 10, 11, 68, 72, 80
Sha Tau Kok   6, 7, 9, 10, 11, 68, 70
Sham Chung   6, 7, 69, 75
Sham Wat   6, 7, 68, 72
Sheung Pak Nai   6, 7, 9, 10, 11, 68, 71, 81
Shui Hau   6, 7, 68, 72
Tai Ho Wan   6, 7, 9, 10, 11, 68, 72
Tai O   6, 7, 68, 72
Tai Tam   6, 7, 13, 68, 73
Tai Tan   6, 7, 9, 10, 11, 69, 74
Tai Wan   6, 7, 9, 10, 11, 69, 75, 78
Tan Ka Wan   6, 7, 69, 74
Ting Kok   6, 7, 9, 10, 11, 69, 73, 79
To Kwa Peng   6, 7, 9, 10, 11, 69, 74
Tolo Pond   6, 7, 9, 10, 11, 12, 69, 73
Tsim Bei Tsui   6, 7, 9, 10, 11, 68, 71
Tung Chung   6, 7, 68, 72
Wong Chuk Wan   6, 7, 69, 75, 78
Wong Yi Chau   6, 7, 9, 10, 11, 69, 74, 78
Yi O   6, 7, 9, 10, 11, 68, 72, 80
Yuen Long Industrial Estate   6, 7, 68, 71